CAMBRIDGE LIBRARY COLLECTION

Books of enduring scholarly value

Physical Sciences

From ancient times, humans have tried to understand the workings of the world around them. The roots of modern physical science go back to the very earliest mechanical devices such as levers and rollers, the mixing of paints and dyes, and the importance of the heavenly bodies in early religious observance and navigation. The physical sciences as we know them today began to emerge as independent academic subjects during the early modern period, in the work of Newton and other 'natural philosophers', and numerous sub-disciplines developed during the centuries that followed. This part of the Cambridge Library Collection is devoted to landmark publications in this area which will be of interest to historians of science concerned with individual scientists, particular discoveries, and advances in scientific method, or with the establishment and development of scientific institutions around the world.

The Scientific Papers of William Parsons, Third Earl of Rosse 1800–1867

William Parsons (1800–67), third Earl of Rosse, was responsible for building in 1845 the largest telescope of his time, nicknamed the 'Leviathan'. It enabled the Earl to make unprecedented astronomical discoveries, including the discovery of the spiral nature of galaxies. Rosse (then Lord Oxmantown) began publishing scientific papers on telescopes in 1828, and for the rest of his life made regular contributions to scientific journals in Ireland, England and Scotland. He served as President of the British Association for the Advancement of Science in 1843, and of the Royal Society from 1848 to 1854, and his addresses to those societies are also included in this collection. Edited by his younger son, the engineer Sir Charles Parsons (1854–1931) and published in 1926, these papers show the wide range of the Earl's interests, from astronomy and telescopes to ancient bronze artefacts and the use of iron in shipbuilding.

Cambridge University Press has long been a pioneer in the reissuing of out-of-print titles from its own backlist, producing digital reprints of books that are still sought after by scholars and students but could not be reprinted economically using traditional technology. The Cambridge Library Collection extends this activity to a wider range of books which are still of importance to researchers and professionals, either for the source material they contain, or as landmarks in the history of their academic discipline.

Drawing from the world-renowned collections in the Cambridge University Library, and guided by the advice of experts in each subject area, Cambridge University Press is using state-of-the-art scanning machines in its own Printing House to capture the content of each book selected for inclusion. The files are processed to give a consistently clear, crisp image, and the books finished to the high quality standard for which the Press is recognised around the world. The latest print-on-demand technology ensures that the books will remain available indefinitely, and that orders for single or multiple copies can quickly be supplied.

The Cambridge Library Collection will bring back to life books of enduring scholarly value (including out-of-copyright works originally issued by other publishers) across a wide range of disciplines in the humanities and social sciences and in science and technology.

The Scientific Papers of William Parsons, Third Earl of Rosse 1800–1867

EDITED BY CHARLES PARSONS

CAMBRIDGE
UNIVERSITY PRESS

CAMBRIDGE UNIVERSITY PRESS

Cambridge, New York, Melbourne, Madrid, Cape Town,
Singapore, São Paolo, Delhi, Tokyo, Mexico City

Published in the United States of America by Cambridge University Press, New York

www.cambridge.org
Information on this title: www.cambridge.org/9781108038072

© in this compilation Cambridge University Press 2011

This edition first published 1926
This digitally printed version 2011

ISBN 978-1-108-03807-2 Paperback

THE
SCIENTIFIC PAPERS
of William Parsons

THIRD
EARL OF ROSSE
1800-1867

COLLECTED AND REPUBLISHED BY

THE HON. SIR CHARLES PARSONS, K.C.B., F.R.S.

1926

PRINTED BY
PERCY LUND, HUMPHRIES & CO. LTD.
THE COUNTRY PRESS   BRADFORD
AND 3 AMEN CORNER, LONDON, E.C.4
BY
LETTERPRESS   AND   THE   L.H.   REPRODUCTION
PROCESS

CONTENTS

INTRODUCTION

WILLIAM PARSONS, third earl of Rosse, F.R.S. (1800-1867) began his experiments for the improvement of reflecting telescopes in 1827, and completed the famous great telescope at Parsonstown, Ireland, with the six-foot speculum, in 1845. His published work concerning the preparation of specula and the construction of telescopes, and the observations (especially on the great nebula of Orion) made at Parsonstown, have not previously been collected from the scientific periodicals in which they appeared. All are included in the present volume, excepting an occasional brief reference which merely duplicates matter more fully dealt with elsewhere. Among the collaborators who aided him in his work of observation, the Rev. Thomas Romney Robinson, F.R.S. (1792-1882) was most closely associated with him. Robinson, distinguished both as astronomer and as ecclesiastic, was director of Armagh Observatory from 1823 until his death, and was president of the Royal Irish Academy from 1851 to 1856. As Rosse contributed nothing to the proceedings of the Academy under his own name, but clearly co-operated closely in the subjects of Robinson's communications on the work at Parsonstown, it has been felt appropriate to include these. Both were men of wide interests outside the range of astronomy, and an example of this is found in the account of the ancient bronze vessel from Rosse's collection, addressed to the Royal Irish Academy by Robinson. Rosse's own correspondence on the construction of ironclads, collected by the Institution of Naval Architects and here reproduced, adds testimony to his inventive genius. His son, the fourth Earl of Rosse (1840-1908), shared his astronomical work and carried it on after his death. The paper on the equatoreal clock, and the account of the observations on the great nebula in Orion (1867) reprinted here, afford evidence of their collaboration ; the fourth Earl published an elaborate résumé of the observations on nebulæ and clusters of stars over the whole period 1848-1878, in the Scientific Transactions of the Royal Dublin Society, 1879. Early references to lunar work, which also was more fully developed after the death of the third Earl, will be found in the present volume.

Rosse was President of the British Association in 1843, and of the Royal Society from 1848 to 1854. His presidential addresses to both bodies are included here.

The papers in this collection, for convenience in composition, have been grouped according to the periodicals in which they originally appeared ; but their chronological order will be readily followed from the contents-pages.

The co-operation of the Royal Society, the Royal Astronomical Society, and the Institution of Naval Architects, in affording facilities for the reprinting of papers from their publications, is gratefully acknowledged.

From the
EDINBURGH JOURNAL OF SCIENCE

Account of a New Reflecting Telescope.

By The Rt. Hon. Lord Oxmantown, M.P., &c.[1]

THE following considerations induced me to make the experiments on the specula of reflecting telescopes which I am about to describe. The reflecting telescope would be almost a perfect instrument, could we devise means of freeing it from spherical aberration ; it would then retain merely the defects necessarily arising from imperfections in the workmanship, and perhaps some others of a much more trifling nature, such as those derived from the inflection of light. The refractor, however, is not only affected by the spherical aberration in common with the reflector, but also by the different refrangibility of the rays of light. Both of these defects may indeed be in a great degree corrected by giving curves of proper radii to the lenses which compose the object-glass. The spherical aberration may by this be almost entirely obliterated, but a considerable portion of the chromatic aberration still remains, owing to the irrationality of the spectra formed by the different kinds of glass, of which the object-glass is necessarily composed, the different coloured rays not being refracted by each in the same proportion. The refractors until lately were limited to a very small scale, owing to the impossibility of procuring suitable glass of large dimensions ; and although a new process has lately been discovered on the continent, which has considerably extended the limits of their construction, still I believe that large pieces of glass, of a tolerably homogeneous nature, are procured with great difficulty ; and there seems to be but little prospect of our being able, with the present state of our knowledge, to construct efficient refractors, at least with glass lenses of apertures at all approaching the late Sir William Herschel's reflectors. I have been thus minute in stating my reasons for making the following experiments, as many practical men whom I have spoken to seem to think that since Fraunhofer's discoveries, the refractor has entirely superseded the reflector, and that all attempts to improve the latter instrument are useless.

[1] Edinb. Journal Sc. IX, No. 17, p. 25 (1828). The name is that of Rosse previously to his succession to the earldom.

Two modes have been hitherto adopted for diminishing the spherical aberration in reflectors, the one by rubbing down the outer surface of the speculum from the edge to the centre, so as to make its figure approach to that of a paraboloid, the other by increasing the focal length in proportion to the aperture. It is certainly extremely probable that a very skilful workman, who has devoted the greater part of his life to the construction of reflectors, may succeed in some instances, particularly when the instrument is what is technically termed a dumpy, in forming a surface approaching to the paraboloid, which will perform better than one which is truly spherical ; but when we consider the extreme accuracy necessary, and that a true surface can only be obtained by the process of polishing, when two motions are combined, the one in some degree at right angles to the other, and that a spherical surface is the only surface which can be formed by these two motions,* it will be evident that when we attempt the parabolic form we abandon an essential requisite to the formation of an accurate surface. It is scarcely worth while remarking, that in every attempt to improve a speculum of an accurate spherical figure I have invariably rendered it worse ;—these attempts were not on very dumpy instruments. The other method of diminishing the spherical aberration by increasing the focal length in proportion to the aperture is certainly unexceptionable ; but it will be immediately evident that it has its limits, and that instruments become unwieldy after they exceed a certain length. I will now immediately proceed to describe one of the instruments I have constructed, with a view of diminishing the spherical aberration without introducing either of these defects.

In Fig. 1, Plate 1, AB is a brass plate turned true on both sides by means of a slide apparatus, which at the same time renders its sides parallel. The dimensions are seven inches by five-eighths thick. CD is another brass plate, made true by the same means, one-half an inch thick. The two plates were then screwed together in a temporary manner, their centres coinciding ; three holes were then bored through them one-fourth of an inch diameter, accurately perpendicular to their surfaces. The two plates were then unscrewed, and the holes in the plate CD were carefully tapped with a tap, having one-sixtieth of an inch interval between its threads. Three cast steel spindles were then accurately turned, the shank EF being made to fit the holes of the plate AB, and the screw GH nicely to fit the holes in the plate CD. It is almost needless to observe that the flanches, and indeed every part of the spindles, must be very carefully turned. The three cast steel spindles were then put through the holes in the plate AB, till their flanches rested on it. They were secured there by washers IK, put on the shanks of the spindles at the back of the plate, and the washers were retained in their places by milled nuts LM. To prevent the washers from shaking as the steel spindles were turned backwards and forwards, which would loosen the milled nuts, each washer was provided with a screw in its side O, which enters the groove P in the shank, and keeps it steady. The plate CD was then laid upon the three screws

* A plane is a spherical surface with an infinite radius, or in practice with a very great radius, and is extremely difficult to execute.

GH of the spindles, which were then gradually turned round in succession by a key fitting their other ends till the plate CD reached within about the one-eighth of an inch of their flanches. To prevent the spindles from shaking either in the plate AB, or CD, lateral holes were drilled reaching the principal holes *o, o, o, v, v, v*. These were stuffed with small bits of leather, which were kept constantly pressed against the spindles by small screws. This precaution is essential. Besides, screws were inserted at the back of the plate through the holes marked *i i i*, which were screwed against the plate CD, to prevent the possibility of its shaking in the slightest degree during the operation of grinding and polishing the speculum.

The mechanism being now complete, a speculum was cast one inch thick. This was secured to the plate CD by three small screwed wires cast into it, by a groove in the plate CD, and by a cement composed of resin, wax, and sulphate of lime. A ring of speculum metal was also cast one inch and a-half thick, which was secured to the plate AB in a similar manner, leaving a very minute interval between it and the piece of speculum metal it surrounded. The whole together formed a speculum of six inches aperture, and two feet focal length, which was ground and polished in the common way, till by repeated trials it was found to be of a good spherical figure. The small screws *i i i*, were then drawn back, the speculum placed in the tube, and the spindles turned round a certain number of times, by which the centre part of the speculum was made to approach nearer the plate AB, by a quantity ascertained in a manner I shall presently point out. The two images were then made to coincide, and the image was then found to be apparently as distinct as either image had been when separate. It is necessary to observe, that, in order to effect the adjustment properly, each image must be brought to the same degree of brightness, which can be accomplished by shades on the mouth of the telescope, and that no higher power should be used than each metal, when separately employed, can bear with distinctness.

I rather think that instruments of this construction will pretty frequently require adjustment ; however, this is easily effected in the space of two or three minutes. The first I attempted consisted of a solid metal surrounded by two rings. Owing to a defect in the mechanism it required very frequent adjustments, the smallest shock displacing the images. The one I have described is almost entirely free from this defect, remaining in perfect adjustment even after very violent shocks. I have a speculum like the first I made, consisting of three parts, which is almost ready for grinding ; and I expect it will turn out well. I have not yet perceived any ill effects from expansion and contraction, which was the difficulty I most apprehended. Whether they will become perceptible in instruments of higher powers, or whether, if perceived, means may or may not be devised of obviating them, can only be ascertained by future trials. On my return from Parliament, if other avocations do not interfere, I propose to construct a speculum in three parts of eighteen inches aperture and twelve feet focal length—this will be giving the experiment a fair trial on a large scale. It may perhaps be as well to observe, that I do not think the principle of subdividing the aberration can be applied with advantage to small instruments. The object to be gained by it is

B

a diminution of the focal length with a given aperture and power, and this is **by** no means desirable in small instruments, as it forces us to make use of deep eye-glasses, which are on many accounts objectionable.

To compute the respective lengths of the curves composing a compound speculum such as has been described, so that the aberration of the whole speculum may be equally divided between them.

Let E, Fig. 2, be the centre of the surface. Let a ray proceed from Q and intersect the axis in V   Let q be the geometrical focus of rays proceeding from the point Q.

Let q represent E Q,   Let $q`$ represent E v,
q' E q, θ R E A
f E F, v ver sin θ

Then considering $q`$ a function of ver. sin θ, v being of course $= 0$ in each coefficient, we have

$$q` = q' + \left(\frac{d\,q`}{d\,v}\right) v + \frac{1}{1.2} \left(\frac{d^2\,q`}{d\,v^2}\right) v^2 + \&c.$$

which by proper substitutions becomes

$$* \; q` = \frac{q\,f}{q+f} + \frac{q^2\,f\,v}{(q+f)^2} + \frac{q^3\,f\,v^2}{(q+f)^3} + \&c.$$

The aberration being $q`-q'$ is evidently represented by this series without its first term, which expressed geometrically amounts to

$$\frac{Q\,E^2}{Q\,F^2}\,\frac{A\,N}{2} + \frac{Q\,E^3}{Q\,F^3}\,\frac{A\,N^2}{4\,E\,F} + \&c.$$

which, when the rays are parallel, becomes

$$\frac{A\,N}{2} + \frac{A\,N^2}{4\,E\,F} \; \&c.$$

It is evident that the first term of this series will afford a sufficiently near approximation for compound specula ; if we therefore represent the ver. sines of the arcs DO, DP, DQ. Fig. 3, by AN, AN', AN'', the problem becomes, Given AN = AN' — AN = AN'' — AN', the length of the arc DQ and its radius, to find the lengths of the arcs DO and DP.

Example.—Let a be the arc DQ = 3 inches, r its radius = 48 inches, x the seconds in a, then † $x = 206265\frac{a}{r} = 206265\frac{1}{16} = 12891''$, which in degrees amounts to 3° 35′ whose v. s. = .0019550.

$\frac{1}{3}$ of which = .0006516, corresponding to arc, 2° 4′ = 7440″ = 1.77 inches.
$\frac{2}{3}$ of which = .0013034 corresponding to arc 2° 54′ 30″ = 10470 = 2.44 inches.
Arc DO = 1.77 inches, and DP = 2.44 inches.

It is evident that the arc PO must be drawn back, a quantity equal to D′C′ = $\frac{1}{2}$ DC. and DO must be drawn back 2 D′C′ = DC.

* Coddington's Optics.   † Lardner's Trigonometry.

Account of an Apparatus for Grinding and Polishing the Specula of Reflecting Telescopes.

By The Rt. Hon. Lord Oxmantown, M.P., &c.[1]

IN Plate IV Fig. 2, AB represents the bed of a large power lathe, CD a shaft connected with the steam engine, EF the floor of the room, GH the ceiling, and IK a large lathe head for very heavy work. Motion is communicated by a band from the shaft CD to the lathe IK, and by bands through a succession of wheels, as represented in the figure to the speculum LM. LM, therefore, revolves with a motion slower than VW, in the proportion of the product of the radii of the wheels transmitting the motion of the product of the radii of the wheels receiving it.

NO represents either the polishing or the grinding tool. PQ is a rod resting against a limber spring RS, and passing through the flat rod TU into a hole in the tool NO. The rod TU is furnished with a joint X, so that when the eccentric VW revolves, the rod XU reciprocates steadily through the guides Y,Y, carrying with it the rod PQ, and of course the tool NO. From S a weight hangs proportioned to the size of the speculum or glass, as the tool alone would be too light.

To make use of this apparatus the speculum LM is secured to the centre of the chuck, as represented in the figure, by means of very soft pitch and wooden pegs driven into the chuck. The tool NO of lead and tin of the proper curve is then placed upon it, and the rods PQ and TU arranged as in the plate. At the commencement of the operation, to expedite the process, the speculum may be made to revolve with considerable rapidity by altering the arrangement of the bands communicating the motion; and the tool NO, for the same purpose, may be prevented from turning by a pin screwed into it, which catches against the rod TU. Coarse emery is now introduced between the tool and speculum through a hole in the tool; and the reciprocating motion of NO, combined with the rotatory motion of LM, will soon bring the speculum to a good surface, though not truly spherical. The focal length of NO being now a little changed, it is replaced by another tool composed of lead and tin, and NO is now suffered to revolve with the speculum, the pin that prevented it being removed; besides the bands are now arranged as in the figure, to give a slow motion to the speculum, and emery of different degrees of fineness is made use of, till the surface of the speculum is extremely smooth.

No fresh emery should now be added for at least a quarter of an hour; and the speed of the shaft CD, and of course of the whole apparatus, is reduced during that time one-half. This seems favourable to the production of a very accurate spherical surface.

* Edinb. Journal Sc. IX., No. 18, p. 213 (1828).   For Figs., see plate facing p. 12.

The speculum will then be ready to be polished, for which purpose the tool is to be covered in the usual manner with a coat of pitch of the proper consistency, and the usual polishing powder applied. At the commencement the shaft CD may revolve with its ordinary speed, and a considerable weight may be applied to the spring RS; this will almost immediately bring the pitch exactly to the same curve as the speculum, and the polish will rapidly proceed; however, towards the end of the operation both the speed of the apparatus and the weight must be greatly diminished.

It appears, then, that the friction by which the polishing and the latter part of the grinding is effected arises from the reciprocating motion of the tool. The circular motion of the speculum is merely for the purpose of continually altering the direction of the strokes resulting from the reciprocating motion. The tool will not revolve exactly in the same time as the speculum; this is to a certain extent useful; but would be prejudicial if friction were produced by it amounting to any sensible quantity. It is principally for the purpose of obviating this that the speculum is made to revolve so slowly. I have tried the introduction of another motion into the process by fixing the guides YY on eccentrics revolving very slowly, and having a very small eccentricity. This seems to be an improvement; but as it renders the machine more complicated, and as I have not yet tried it sufficiently to be able to speak with certainty of its good effect, I have not represented it in the figure.

It is evident that a considerable number of specula or lenses may be worked at the same time, by increasing the number of the wheels Z. I have not, however, extended the apparatus beyond what the figure represents, having no occasion for it.

The annexed sketch has no pretentions to the accuracy of a working drawing. The scale is merely for the purpose of giving a general idea of the proportions of the parts. The machine would be sufficiently large to grind and polish a speculum of three feet diameter or perhaps larger. Should an instrument-maker think it worth his while to construct an apparatus on the above principle, probably one on one-third of the scale would be sufficient for his purposes;—it might be moved by men and a porter's wheel. The quantity of power necessary would of course depend upon the velocity of the machine, and the size and number of the specula to be worked.

The engine I make use of in my laboratory is a two horse power; and from some loose experiments with it, I should think that a one horse power would be fully sufficient for three or four specula six inches diameter. The rod TU may make a stroke of about an inch for a speculum of the above dimensions, and the revolutions of the speculum may be to the revolution of the eccentric as 1 : 300 or 400. For specula of six inches diameter a day will be found sufficient to complete the process.

It is evident that the object of this apparatus is to communicate an accurate spherical figure. Should it be proposed to attempt the parabolic curve, the motion recommended for that purpose might be imitated by means of the eccentric guides and the slow circular motion of the speculum; and with this advantage, that, were it found really successful, the same result would probably be always afterwards obtained; but, from reasons which I have stated in an account of an attempt to improve the reflector (see last Number, p. 25)[1], I fear the parabolic curve is inconsistent with an accurate surface.

In Fig. 3, AB represents a lathe for cutting the tools for grinding specula and lenses. The mandril works between brass jaws secured by a variety of screw bolts to prevent any shake. It is more to be depended upon than one of the common construction, and admits of easier adjustment. CD performs the office of the slide puppet of a common lathe, but is steadier, and in other respects preferable to it. The triangular bar EF is moved backwards and forwards by the screw G. IH is a light frame of wood. K the cutting edge. IM a rod of iron made flat at I and pierced with a small hole. The rod IM may be drawn out more or less; and then it can be secured by a screw.

To cut a convex tool, the lathe AB and puppet CD are placed upon the bed of the lathe, Fig. 2, A and F being turned towards each other. The axis of the mandril AB and of the triangular bar CD are then brought into the same right line, and placed at the proper distance. The frame IH is then adjusted so that the interval between the hole at I and K shall be exactly equal to the radius of the tool to be cut. The extremity of the bar IM is then inserted into the horizontal slit in the bar EF, and then secured by a steel pin, upon which it plays, and the edge K is carried by a slide rest along the surface of the tool, which had been previously fixed on at B, and in a horizontal line corresponding with its diameter; and the impression it makes may be regulated by the screw C. A very accurate spherical surface of exactly the required radius will thus be formed. A concave tool is cut in the same manner, only the lathe AB is reversed, and a simple radius is substituted for the frame IH.

I have cut tools with this of twelve feet radii, but shorter radii are more manageable.

The whole of the above apparatus has been very lately constructed. When it has been more used, improvements will no doubt suggest themselves. The interposition of another wheel and spindle between VW and LM would be of use, as it is rather difficult without it to communicate a sufficiently slow motion to the speculum; but upon the whole, I think this machinery will be found greatly to abridge the labour and increase the certainty of the process of working specula and lenses.

[1] P. 2 of this volume.

Account of a Series of Experiments on the Construction of Large Reflecting Telescopes.

By The Rt. Hon. Lord Oxmantown, M.P., &c.[1]

HAVING, at different intervals during the last three years, tried a variety of experiments on the construction of specula for large reflecting telescopes, perhaps some of the results which I have arrived at may not be uninteresting to the scientific public.

In making these experiments, I have had two objects in view, *first*, to ascertain whether it was practicable to remove any of the defects known to exist in the large reflecting telescopes hitherto constructed ; and, *secondly*, to simplify the process necessary for the manufacture of good reflecting telescopes of ordinary dimensions, so that the art might be no longer a mystery, known to but few individuals, and not to be acquired, but after many years of laborious apprenticeship.

A general statement of the results of my experiments will enable those who are at all conversant with the use of telescopes to decide how far I may have succeeded in effecting anything useful.

I propose to avoid as much as possible entering into detail. Within the limits necessarily prescribed for a single article in a periodical work, it would be impossible to do so with any advantage. In subsequent numbers of this *Journal*, I shall have an opportunity of giving a particular account of the different processes and manipulations which I have employed, so that any person of ordinary mechanical skill who may think it worth while to erect the necessary machinery, will be enabled to obtain with certainty the same results.

As a general inference from all the facts which have come within my observation, I can have no hesitation in stating, that the reflecting telescope is still susceptible of very great improvement,—that it has by no means reached the utmost limits of perfection. If we except the defects arising from spherical aberration and the inflection of light, which I think are not irremediable, and are, in my opinion, much overrated in practice, the remaining defects are entirely of a practical nature, and to be overcome by practical means, by numerous and accurate experiments, such as a patient consideration of the difficulties to be surmounted must necessarily suggest.

In order to render the following account intelligible, I will endeavour to put the reader in possession of the difficulties he would have to encounter were he to proceed to construct a large telescope in the common way, and the defects he would probably find in the instrument when finished. He would of course first proceed to cast the metal. As earthen vessels would not be sufficiently capacious, he would employ either iron ones or a reverberating furnace. If he tried iron vessels, before

[1] Edinburgh Journal of Science, n.s. II, 136 (1830).

a large quantity of speculum metal, for instance three or four hundred weight, was raised to a proper heat for casting, he would find that the metal had imbibed some of the iron, and was injured ; or perhaps, if he was less fortunate, and the fire had been a little mismanaged, that the speculum metal had promoted the fusion of the iron, and so passed out through the crucible. The reverberatory furnace would then be resorted to. Much difficulty would occur in combating the continual change of the quality of the metal from the exposure of so large a surface to the action of the flame. However, the metal once cast, the next process would be to anneal it. He would then find that the speculum would fly to pieces before it was cool, unless the alloy made use of was less bright, less white, and in every respect inferior to the best speculum metal. The next process is to grind the speculum, which, though laborious, does not require much exactness, and lastly, to polish it, which every one knows is attended with very great difficulty. Making a probable estimate of the success likely to be obtained, after a great number of abortive attempts, a metal would be completed, having a tinge of yellow deeper in proportion to its size, with perhaps a defective polish, and certainly a figure by no means perfect ; such a metal would not bear any considerable power with tolerable distinctness. What I have just stated is the result of experience. That I have not overrated the difficulties and defects, will appear evident to any one who is conversant with the late Sir W. Herschel's writings. Since Sir W. Herschel's time, no improvement that I am aware of has been made in any part of the process of making the specula of telescopes ; none of the difficulties which he stated as existing have since been surmounted ; and none of the defects which his skill had not removed, have since yielded to the dexterity and perseverance of others.

From the accounts which we have of Sir W. Herschel's labours, it appears, that, in proportion as he increased the size of his specula, he was obliged to debase the quality of the metal made use of. Dr Pearson, in his *Practical Astronomy*, states, that the proportion of tin to copper used for the metal of the twenty foot telescope was 7.75 to 20,—an alloy certainly extremely low. He also states that the metal of the forty foot telescope was still lower, and was composed of blocks of an alloy purchased at a warehouse in London. The weight of the large metal for the forty foot telescope was 1050 pounds, the diameter four feet eight inches, the thickness at the edge two inches, and in the middle one inch and a quarter. It is difficult to conceive how a metal of such weight, so great a diameter, and so little thickness, could retain even a tolerable figure in the different positions of the telescope.

It is also well known, that Sir William Herschel, at the commencement of his career, polished 400 specula of different dimensions ; content if he could procure one tolerably good one out of a great number. Such were the difficulties that he had to encounter ; and I am not aware that anything has been published since that time, tending materially to diminish the labours of the experimentalist, or of the practical optician. Sir William Herschel also found that he was unable to polish large specula so as to give them as accurate a figure as small ones. His twenty foot telescope, which I believe has been admitted to be the best reflecting telescope ever constructed, was seldom used with a power above 200 ; and I believe the same observation will apply with equal correctness to the forty foot telescope.

The defects, therefore, common to all very large specula hitherto constructed may be thus stated : a defective metallic composition ill suited either to receive or retain a polish, or to show objects of their natural colour and brilliancy ; a want of sufficient stiffness in proportion to their weight to enable them to retain their figure with that great degree of exactness necessary ; and *thirdly*, a want of as perfect a polish and figure as has been given to small specula.

My first experiments were undertaken with the view of obviating the two first defects. Having had some experience in the process of painting on glass, in which the glass is made red hot, and subsequently annealed, it occurred to me that the precautions employed in that process might be transferred with advantage to the construction of specula, and that it might thus be practicable to prevent large specula, cast of the highest metal, from cracking before they were finished. It also occurred to me that large specula might possess sufficient stiffness without any additional weight, were they cast thin, but with a deep rim round them connected by ribs of equal depth. A speculum, fifteen inches diameter, was accordingly cast with a rim round the edge two and a half inches deep, and half an inch thick, and with two ribs of the same depth and thickness as the rim, intersecting each other at the centre of the back of the speculum. The composition employed was the best speculum metal. As soon as the metal had become solid, while still red hot, the sand was entirely removed from the four cavities at the back between the ribs and the rim, and the metal, still red hot, was placed in a red hot iron vessel upon a bed of wood ashes, and the cover of the vessel, also red hot, was then put on, the whole was immediately placed in a red hot oven and shut up there. In about forty-eight hours the metal was perfectly cool. It was then examined, and was found to be broken in several places. A second metal was then cast precisely similar to the former one and similarly treated, but the composition was a little lowered. It also cracked, but not so much. A third was cast, the composition being a little lower than that of the second ; it also met with a similar fate. A fourth of a still lower composition was defective in casting, but did not crack ; the fifth turned out well. The fourth and fifth were of the same composition. The metal has a slight tinge of yellow clearly perceptible when compared with metal of the best composition. It does not take so high a polish, and is more subject to tarnish. The metal, however, was much higher than that of Sir William Herschel's twenty foot telescope.

Upon the whole, the result of the above-mentioned experiments was by no means satisfactory. I found that I could not cast a speculum of the moderate dimensions of fifteen inches, without reducing the composition considerably below the highest standard. It was also quite evident that the composition should be still lower for a metal three feet diameter ; such a metal might indeed have been made of one-third the weight which would otherwise have been necessary, by casting it like the fifteen inch metal with a rim and ribs at the back ; but still the defect in the quality of the metal would have remained, which appeared to me to be a decisive objection to the construction of such an instrument.

After several fruitless attempts to combat this difficulty, experiments were tried to ascertain whether it would be practicable to cast specula in different pieces, and to unite them together by tinning the surfaces. This was found to be practicable, but it was abandoned for the following plan, which I think was perfectly successful.

An alloy of zinc and copper can be formed, which will expand and contract with changes of temperature more or less than speculum metal, according to the proportion of the ingredients. Experiments were made, and it was found that copper $2+\frac{3}{4}$, and zinc 1, would give an alloy possessing the required property of giving expansions and contractions with a change of temperature, not sensibly different from speculum metal. This alloy is malleable, ductile, and easily worked. With this alloy a speculum was cast fifteen inches diameter, with a rim and ribs similar to the one before described, but in every respect thinner,—not half its weight. It was turned smooth and flat at one side and tinned. Six pieces of the highest speculum metal were then prepared one quarter of an inch thick, and fitted so as to make, when put together, a complete circular disc fifteen inches diameter ; these were then arranged on the flat tinned surface of the brass speculum ; the temperature was then very gradually and equally raised till the tin was in fusion, and till every part of the under sides and edges of the speculum metal was perfectly tinned. A slight pressure was then uniformly applied, and the temperature gradually reduced till the tin became solid. We then had a speculum composed of zinc and copper plated with speculum metal one quarter of an inch thick, adhering to it as firmly in every part as if it had been one piece of metal. This metal was ground and polished by the machine described in a former number of this *Journal*. It has a focal length of twelve feet, and as there is a set off of about a quarter of an inch at the edge, it has fourteen inches and a half clear aperture. It far surpasses the other metal in the brilliancy and whiteness of the image, as was of course to be expected. In other respects it is the same, as they both bear distinctly a power of 600 at a printed paper, or at the cut stone pinnacles of a church distant about 300 yards. There can be no mistake as to the powers, as I make use of single lenses. Only one favourable night has occurred since the stand and its appendages have been brought to such a state as to render the instrument tolerably manageable. The new moon was on that occasion examined with powers from 80 to 600, and very perfectly defined. The pole star, ε Bootis and some other stars not requiring high powers were well shown. From the defective state of the machinery for giving motion to the telescope high powers were not employed, and farther trials were deferred till another opportunity. The stand is precisely similar to Mr. Ramage's, but the pulleys and some other appendages are not complete. From a comparison of the instrument in the day time with others which perform extremely well, I entertain very sanguine expectations of its powers upon the difficult double stars. A six inch metal was constructed about a year ago upon the same principle, which performed well ; and, upon the whole, I have not been able to discover that these plated specula are subject to any defects to which those upon the common construction are not equally liable.

It is evident that such specula can be constructed of the finest metal and of any size which may be desired, and that with the greatest facility. A metal upon this plan two feet diameter was commenced a few days ago, and is in so forward a state that it will probably be completed in three weeks. A second speculum of the same dimensions could be completed in a shorter time, as no fresh tools would be necessary.

This metal is for a tube twenty-six feet long and three feet diameter. The tube is finished and the stand is nearly so. I propose to make another metal for it at a future time of the full aperture ; but whether single, or upon the plan described in Number 17 of this *Journal*[1], I have not yet determined. Further experiments are to be tried with the six inch metal, the subject of the article before quoted. A few months after that article was sent to the press, an eighteen inch metal of twelve feet focus upon the same plan was commenced ; the different parts were cast. In the meantime some further experiments were tried with the six inch metal ; the power of adjustment afforded much facility for comparing the spherical aberration with the defects proceeding from other causes. By retaining but a section of the aperture the spherical aberration was preserved, while the defects arising from inaccuracy of surface were reduced with the diminished aperture. The distinctness of the images increased as the surface lessened, and the image resulting from the union of the two images was as distinct as each had been when separately examined. Upon the whole, it appeared evident, that, although the speculum was improved by the adjustment for spherical aberration, still defects continued, arising from the imperfections of its surface, much greater than the spherical aberration. The speculum was repolished by hand with the utmost care more than fifteen times, but without any considerable improvement. It was compared when taken from the polisher with a common speculum of the same dimensions, and they were found to be both alike. The compound one when adjusted was somewhat superior. The polish of both was very good, but the surface of course was not so. The solid speculum was about as good as the average of similar metals which I have seen. When these metals have been ground and polished by machinery, further trials shall be made with them.

After the experiments which I have just described, I determined to defer for the present expending any further labour upon the eighteen inch compound speculum, and resolved to endeavour previously to discover some certain method of giving specula a more accurate surface. I was confirmed in that determination from an apprehension that the castings were not as perfect as they should have been. All my workmen were trained in my own laboratory without the assistance of any professional person, and none of them had previously seen any process in the mechanic arts ; and I was not myself then acquainted with the precautions necessary to insure the production of an alloy of zinc and copper in the due proportions. There was, therefore, a great probability that the castings were defective.

The polishing apparatus described in No. xviii. of this *Journal*[2] was completed about that time. It has since undergone some alterations. The different motions

[1] Pp. 1 *seq.* of the present collection.
[2] Pp. 5 *seq.* of the present collection.

PLATE I

Edin.ʳ Jonrᵗ of Science Vol. IX

Fig.1

Fig 2

Fig 3

PLATE IV

Edin. Jourᵗ of Science Vol. IX

Fig 2

1   2   3   4   5   6   7   8   9   10   11   12   13   14   15   16   Feet

Fig. 3

are now obtained by cogwheels and leather bands. Several other minor alterations have also been made, both in the apparatus and in the manner of conducting the whole process, which have produced the most material improvement. The results obtained by machinery are very nearly uniform. Where a uniform combination of motions produces a defect, that defect will uniformly recur, and may, therefore, with great facility, be traced to its source and corrected. Such has repeatedly been the case. The same specula have been repolished a great number of times, and the performance of the machine has improved faster than I could have anticipated.

The practical optician will rarely give you the slightest intimation of the process of working specula which he finds the most successful, nor is it perhaps to be expected. It is therefore impossible to describe with certainty his mode of proceeding ; but I believe the practice is to work the speculum till it becomes warm, and the polisher is almost dry ; and I rather think that practical opticians suppose it is impossible to communicate a fine polish without this mode of proceeding. From the experiments which I have tried, I have little doubt but that the figure of the speculum is injured by working it upon a polisher nearly dry, and that the injury is in some degree proportional to the time it is so worked :—at any rate large metals must be finished upon a moist polisher. Until very lately I had not found out a method of communicating a very fine polish to a cold metal worked upon a moist polisher.

As fine a polish as can be desired can now be given to a metal of any temperature which we may fix upon above the freezing point. Both theory and practice lead to the same conclusion, that it is desirable to polish a speculum at a temperature as nearly as possible the same as that at which it is to be afterwards used, particularly if the speculum is of large dimensions.

In the preceding account, I have endeavoured to give a general outline of the different objects which I have attempted to effect, and I have, as far as was in my power, conveyed an accurate idea of the degree in which I conceive I have been successful. Further experiments shall be tried, and the specula already completed shall be subjected to the severest tests. Should I then feel satisfied that specula obtained by these processes are as perfect as I have ventured to anticipate, I shall then have the pleasure of placing some of them in the hands of the able and persevering observers of the present day, where they will be fairly tried, and, if they have merit, will certainly not remain idle.

The examination of the heavens commenced by the late Sir William Herschel, and, prosecuted by him with such success, still continues. New facts are recorded ; and there can be little doubt but that discoveries will multiply in proportion as the telescope may be improved.

It is perhaps not too much to expect, that the time is not far distant when data will be collected sufficient to afford us some insight into the construction of the material universe.

From the
PROCEEDINGS OF THE ROYAL IRISH ACADEMY

Account of the Three-feet Telescope.

By Dr. T R. Robinson[1]

THE Rev. T. R. Robinson, D.D., M.R.I.A., gave the Academy an account of a large reflecting telescope, lately constructed by Lord Oxmantown, and of the processes employed in forming its specula.

After explaining the relative importance of magnifying and illuminating power, Dr. R. proceeded to give a brief sketch of the history of the reflecting telescope, which seemed to have been forgotten for many years after its invention, till it was revived by Hadley. The labours of Short soon gave it celebrity ; yet even this artist limited himself in almost every instance to sizes which were not more powerful than the achromatics of his day, and his large instruments appear to have been failures.[2] It was not till a full century after the publication of Newton's paper, that Sir William Herschel gave this telescope the gigantic development which has crowned him with imperishable fame ; and by the construction of telescopes of nineteen and forty-eight inches aperture, placed regions almost beyond the scope of measurement within the reach of human intellect. But as Short, in a spirit unworthy of his talents, took care that his knowledge should die with himself, and Herschel published nothing of the means to which his success was owing, the construction of a large reflector is still as much as ever a perilous adventure, in which each individual must grope his way. Accordingly, the London opticians themselves do not like to attempt a mirror even of nine inches diameter, and demand a price for it which shows the uncertainty and difficulty of its execution. In Ireland we are more fortunate, for a member of our Academy, Mr. Grubb, finds no difficulty in making them of admirable quality up to this size, or even fifteen inches ; but with all his distinguished mechanical talent, he is believed to be doubtful of the possibility of more than doubling this last magnitude in perfect speculum metal.

[1] Proceedings R. I. A. II, 2 (1840).

[2] A Newtonian of six feet focus, and 9·4 inches aperture, is said by Maskelyne to have shown the first satellite of Jupiter 13″ longer than a *triple* achromatic of 3·6 inches aperture. The telescope of twelve feet focus, and eighteen inches aperture, now at Oxford, showed multiple rings of Saturn.

Under these circumstances, too much praise cannot be given to Lord Oxmantown, who, in the midst of other pursuits, has found leisure for such researches ; and by a rare combination of optical science, chemical knowledge, and practical mechanics, has given us the power of overcoming the difficulties which arrested our predecessors, and of carrying to an extent which even Herschel himself did not venture to contemplate, the illuminating power of this telescope, along with a sharpness of definition scarcely inferior to that of the achromatic.

The chief difficulties which are to be overcome in the construction of reflectors, arise from the excessive brittleness of the composition of which specula are made, and from the necessity of giving them figures which shall be free from aberration. The great mirror in the Newtonian form is (if the eyepiece and plane mirror be correct) the conical paraboloid.

It is necessary that speculum metal should possess, in the highest attainable degree, the qualities of whiteness, brilliancy, and resistance to tarnish. Lord Oxmantown has found that these conditions are best satisfied in the *definite* combinations of four equivalents of copper to one of tin ; or by weight, 32 and 14·7 nearly. Metals differing from this by a slight excess of either component, are, when first polished, scarcely less brilliant, but are dimmed so rapidly that the lapse of a few days produces a sensible difference. On the other hand, some large specula of the atomic compound have been lying uncovered for years, without material injury to their polish.

But this compound is brittle almost beyond belief ; a slight blow, or even the application of partial warmth, will shiver a large mass of it ; though harder than steel, its surface is broken up with the utmost facility, and it has a most energetic tendency to crystallize. The common process of the founder fails with it, except for masses of very limited magnitude, as the cast cracks in the mould, and the subsequent difficulties of the annealing are such, that it has been a very general practice to use an alloy lower (containing more copper) than the atomic standard. Even Sir William Herschel was obliged to yield to this necessity. It appears from a letter of Smeaton, (Rees' Cyclopædia, Art. Telescope), that for his 20 feet mirror of 19 inches aperture, the composition was 32 copper to 12·4 tin ; and that for the 40 feet it was even lower. Yet two out of these attempts to cast this huge speculum failed.

Lord Oxmantown at first endeavoured to evade the difficulty, by constructing a speculum in pieces, soldering plates of fine metal to a back of a peculiar brass, ascertained to have the same expansion ; and has completed one of thirty-six inches aperture and twenty-seven feet focal length, which performs very well on stars below the fifth magnitude, but above that exhibits a cross formed by the diffraction at the joints ; and in unsteady states of the air exhibits the sixteen divisions of the great mirror on the star's disk. By diminishing the number and size of the joints it is found, that these inconveniences can be diminished, so as to be scarcely perceptible ; and in all probability this is the process by which the remotest limits of telescopic vision will ultimately be attained. It is, however, not necessary for instruments of even greater dimensions than this, since Lord Oxmantown has succeeded, by a contrivance as simple as ingenious, in casting at the first attempt a *solid* mirror of the same size ; and there is no reason to suppose that it will be less effective on a much larger scale.

But however difficult it may be to obtain the rough speculum of large dimensions, it is still more so to give it a proper figure, combined with that brilliant polish which is technically called black, because it reflects no light out of the plane of incidence.   In such mirrors as can be wrought by hand, they are worked by short cross strokes on the polisher, and at the same time have a slow rotation relative to it.   This might be expected to produce merely a spherical figure ; but by varying the length of the stroke, by circular movement, elliptic figure of the polisher, or removing portions of its pitch covering, a parabolic figure is obtained.   For sizes above nine inches diameter, the work must be performed by machinery ; but in all which Dr. R. has seen (the most remarkable of which are those of Sir William Herschel[1] and Mr. Grubb), the cross stroke is given by a lever moved by hand ; and it is supposed that perfect results cannot be obtained but by the *feeling* of the polisher's action.   Sir John Herschel is believed to have made important additions to his father's apparatus ; and it is to be hoped he will soon redeem his promise (Mem. R. Ast. Soc. vol. vi.) of publishing his improvements.

Lord Oxmantown has in many respects deviated from the usual process.   His polisher, of the mirror's diameter, intersected by transverse and circular grooves, into portions not exceeding half an inch of surface, is coated, first, with a thin layer of the common optical pitch, and then with a much harder compound.   It is worked *on* the mirror, and counterpoised so that but little of its weight bears ; but the want of pressure is compensated by a long and rapid stroke.   The mirror revolves slowly in a cistern of water, maintained at a uniform temperature, to prevent the extrication of heat by friction.   The polisher moves slowly in the same direction, while it is also impelled with two rectangular movements.   The machine is driven by steam, and requires no superintendence, except to supply occasionally a little water to the polisher, and to watch when the polish is complete. By an induction from experiments on mirrors from six to thirty-six inches aperture it was found, that if the magnitudes of the transverse movements be $\frac{1}{8}$ and $\frac{9}{100}$ of the aperture, and their times be to its period of rotation as 1 and 1·8 to 37, the figure will be parabolic ; but to combine with this the highest degree of lustre, it is found necessary to apply, towards the close, a solution of soap in liquid ammonia, which seems to exert a specific action.

The certainty of the process is such, that the solid mirror of thirty-six inches aperture, after being scratched all over its surface with coarse putty, was, in Dr. R.'s presence, perfectly polished in about six hours, and was placed in its tube for examination, without any previous trial as to quality.

Lord Oxmantown has preferred the Newtonian to the Herschelian form, and, in Dr. R.'s opinion, with good reason.   In the latter, the inclination of the great

---

[1] Dr. R. had the good fortune to see this at Slough, in 1830, while at work on a twenty-feet mirror.

mirror to the incident rays must deform the image,[1] and it is now known, that even with faint objects, sharp definition is of high importance. It should, in fact, be a segment of a paraboloid, exterior to the axis ; and though a theorem of Sir William Hamilton (Trans. R. Irish Acad., vol. xv. p. 97), might seem to indicate mechanical means of approximating to the figure, yet Dr. R. fears there would be greater difficulty in applying them than in enlarging the aperture of the New-tonian, so as to make up for the loss of light. Another serious objection is, that in the Herschelian the observer's position at the mouth of the tube, must cause currents of heated air, which will materially interfere with sharpness of definition.

As to the loss of light by the second reflexion, Dr. R. thinks it has been much overrated, and expresses a wish that a careful set of experiments were made on reflexion by plane specula at various incidences, on prisms of total reflexion, and the achromatic prism, proposed as a substitute by Sir David Brewster.

As to the rest of the instrument, it may suffice to say, that it bears a general resemblance to that of Ramage, but that the tube, gallery, and vertical axis of the stand are counterpoised, so that one man can easily work it, notwithstanding its enormous bulk. The specula, when not in use, are preserved from moisture or acid vapours, by connecting their boxes with chambers containing quicklime, which is occasionally renewed. This arrangement (which also occurred to Dr. R., and has been for several years applied by him to the Armagh reflector), appears to be very effective in preserving the polish.

In trying the performance of the telescope, Dr. R. had the advantage of the assistance of one of the most celebrated of British astronomers, Sir James South ; but they were unfortunate in respect to weather, as the air was unsteady in almost every instance ; the moonlight was also powerful on most of the nights when they were using it. After midnight, too (when large reflectors act best), the sky, in general, became overcast. The time was from October 29th to November 8th.

Both specula, the divided and the solid, seem exactly parabolic, there being no sensible difference in the focal adjustment of the eyepiece with the whole aper-ture of thirty-six inches, or one of twelve ; in the former case there is more flutter, but apparently no difference in definition, and the eyepiece comes to its place of adjustment very sharply.

The solid speculum showed α Lyræ round and well defined, with powers up to 1000 inclusive, and at moments even with 1600 ; but the air was not fit for so high a power on any telescope. Rigel, two hours from the meridian, with 600, was round, the field quite dark, the companion, separated by more than a diameter of the star from its light, and so brilliant that it would certainly be visible long before sunset.

ζ Orionis, well defined, with all the powers from 200 to 1000, with the latter a wide black separation between the stars ; 32 Orionis and 31 Canis minoris were also well separated.

---

[1] Any one who has a Newtonian telescope can verify this, by inclining a little the great mirror, so however as not to pass the edge of the plane mirror by the pencil. In Lord O.'s instrument, an inclination of 11′ sensibly injures it ; were it Herschelian, the inclination must be 3° 11′.

It is scarcely possible to preserve the necessary sobriety of language, in speaking of the moon's appearance with this instrument, which discovers a multitude of new objects at every point of its surface. Among these may be named a mountainous tract near Ptolemy, every ridge of which is dotted with extremely minute craters, and two black parallel stripes in the bottom of Aristarchus.

The Georgian was the only planet visible ; its disc did not show any trace of a ring. As to its satellites, it is difficult to pronounce whether the luminous points seen near it are satellites or stars, without micrometer measures. On October 29, three such points were seen within a few seconds of the planet, which were not visible on November 5 ; but then two others were to be traced, one of which could not have been overlooked in the first instance, had it been in the same position. If these were satellites, as is not improbable, there would be no *great* difficulty in taking good measurement both of their distance and position.

There could be little doubt of the high illuminating power of such a telescope, yet an example or two may be desirable. Between ϵ^1 and ϵ^2 Lyræ, there are two faint stars, which Sir J. Herschel (Phil. Trans. 1824) calls " debilissima," and which seem to have been, at that time, the only set visible in the twenty-feet reflector. These, at the altitude of 18° were visible *without an eye-glass*, and also when the aperture was contracted to twelve inches. With an aperture of eighteen inches, power 600, they and two other stars (seen in Mr. Cooper's achromatic of 13·2 aperture, and the Armagh reflector of 15) are easily seen. With the whole aperture, a fifth is visible, which Dr. R. had not before noticed. Nov. 5th, strong moonlight.

In the nebula of Orion, the fifth star of the trapezium is easily seen with either speculum, even when the aperture is contracted to eighteen inches. The divided speculum will not show the sixth with the whole aperture, on account of that sort of disintegration of large stars already noticed, but does, in favourable moments, when contracted to eighteen inches. With the solid mirror and whole aperture, it stands out conspicuously under all the powers up to 1000, and even with eighteen inches is not likely to be overlooked.

Comparatively little attention was paid to nebulæ and clusters, from the moonlight, and the superior importance of ascertaining the telescope's defining power. Of the few examined were 13 Messier, in which the central mass of stars was more distinctly separated, and the stars themselves larger than had been anticipated ; the great nebula of Orion and that of Andromeda showed no appearance of resolution, but the small nebula near the latter is clearly resolvable. This is also the case with the ring nebula of Lyra ; indeed, Dr. R. thought it was resolved at its minor axis ; the fainter nebulous matter which fills it is irregularly distributed, having several stripes or wisps in it, and there are four stars near it, besides the one figured by Sir John Herschel, in his catalogue of nebulæ. It is also worthy of notice, that this nebula, instead of that regular outline which he has there given it, is fringed with appendages, branching out into the surrounding space, like those of 13 Messier, and in particular, having prolongations brighter than the others in the direction of the major axis, longer than the ring's breadth. A still greater difference is found in 1 Messier, described by Sir John Herschel, as " a barely resolvable cluster," and drawn, fig. 81, with a fair elliptic boundary. This telescope, however, shows the stars, as in his figure 89, and some more plainly, while the general outline, besides being irregular and fringed with appendages, has a deep bifurcation to the south.

From these and some other discrepancies, Dr. R. thinks it of great importance that the globular nebulæ and clusters should be all carefully reviewed, as it is chiefly from their supposed regularity that the hypothesis of the condensation of nebulous matter into suns and planets has arisen, an hypothesis which he thinks has, in some instances, been carried to an unwarrantable extent.

On the whole, he is of opinion that this is the most powerful telescope that has ever been constructed. So little has been published respecting the performance of Sir W. Herschel's forty-foot telescope, that it is not easy to institute a comparison with *that*, the only one that can fairly be made to compete with it. But there are two facts on record which lead to the inference that it was deficient in defining power ; one, the low power used, which Dr. R. thinks was not above 370 ; the other, the circumstance that neither the fifth nor sixth stars of the trapezium of the nebula of Orion were shown by it. As to light, there is no reason to believe that the composition of the forty-foot mirror was as reflective as that of the twenty-foot ; and if Dr. R. be correct in the opinion, that the latter[1] did not show the fifth star easily, or the sixth at all, and that it only exhibited the " debilissima," and one star near the ring-nebula, then *it* has decidedly less illuminating power than eighteen, perhaps not more than fourteen inches aperture of Lord Oxmantown's mirror, notwithstanding the loss of light in that by the reflexion at the second speculum.

However, any question about this optical pre-eminence is likely soon to be decided, for Lord Oxmantown is about to construct a telescope of unequalled dimensions. He intends it to be six feet aperture, and fifty feet focus, mounted in the meridian, but with a range of about half an hour on each side of it. If he succeeds in giving it the same degree of perfection as that which he has attained in the present instance, which is exceedingly probable, it will be, indeed, a proud achievement ; his character is an assurance that it will be devoted, in the most unreserved manner, to the service of astronomy, while the energy that could accomplish such a triumph, and the liberality that has placed his discoveries in this difficult art within reach of all, may justly be reckoned among the highest distinctions of Ireland.

In its original state, not as improved by the more perfect means latterly employed by Sir John Herschel.

On Lord Rosse's Telescope.

By Dr. T. R. Robinson[1]

DR. Robinson, when giving, in November, 1840, to the Academy, an account of the three-feet telescope constructed by the Earl of Rosse, had announced to them the intention of that nobleman to attempt an instrument of double aperture and focal length. The attempt had succeeded even beyond expectation, and he hoped that a brief notice of its progress and results would not be uninteresting ; more especially as he felt that the approbation with which they had received his former communication, and the importance which they attached to Lord Rosse's discoveries, had not been the least powerful cause of the triumph which their countryman has now achieved.

The speculum was cast on the 13th of April, 1842, according to the principles which had been so successfully applied to the smaller mirrors ; but with several changes of the details, made necessary by the gigantic scale of the work.

It is well known to all who have experimented on specula, that the alloy must be formed in the first instance, and remelted for casting at a much lower heat : otherwise the mirror is full of pores. The fusion must, in both cases, be effected in covered crucibles, to preserve the definite proportions of the alloy, which would be *lowered* by oxidation of its tin if exposed to the draught of the furnace. It is also necessary that the speculum be of uniform composition and superficial density ; and as it is impossible to fuse the requisite quantity of metal for one of six feet in a single vessel, the different portions must flow into the mould under the circumstances as nearly as possible identical. Much thought and many experiments must have been expended before these conditions were so completely fulfilled. The crucibles are, of course, cast iron ; no earthen one being able to bear the pressure of such a mass of fluid metal at so high a temperature. They are thirty inches internal depth, and twenty-four diameter, weighing about half a ton each, and manufactured with the precautions pointed out by Lord Rosse (Phil. Trans. 1840)[2] Notwithstanding their great strength, they yield so much that it is obviously hazardous either to use them frequently or to increase their dimensions. Three were employed at once, each containing about one and a half tons of the alloy : they were placed in furnaces whose mouths were level with the ground, eight feet deep and four in diameter, disposed round a large stack or chimney, into which their flues vent. The fuel is turf, peculiarly fitted for this work, as giving a much more manageable heat than coke ; about 2,200 cubic feet of it are consumed in a casting. The furnaces were filled with fuel, and ignited at the top, on the preceding evening, that the crucibles might be gradually heated ; and in about ten hours

[1] Proceedings R. I. A. III, 114 (1845).
[2] Pp. 80 *seq.* of the present collection.

they were ready to receive the metal. This was unintentionally made of a lower standard than that of the three feet, in consequence of the atomic number for tin being taken as given in Turner's Chemistry, 57.9 instead of 58.9, causing a deficiency of about half per cent., too trifling to impair materially its reflective power, though it will certainly make it more liable to tarnish. That its uniformity might be insured, each ingot of it was broken into three pieces, as nearly equal as possible, and stored in three casks, each of which contributed equally to form the successive charges of the crucibles in an order regularly varying. They were charged at intervals of two hours, and the whole was fused in twelve : they were then withdrawn from the furnaces by a powerful crane, and transported to the iron cradles of pouring frames, arranged 90° asunder round the mould.

The essential part of the mould is its base, composed of hoop iron six inches broad, packed on edge in a strong frame seven feet diameter, and supported by strong transverse bars below. The upper surface was turned to a convex segment of a sphere, 108 feet radius, on the polishing machine, over which a self-acting slide rest was fixed, whose frame was of the same curvature. This process required several weeks, and it was then ground smooth by a frame filled with concave blocks of sandstone. The bed of hoops so prepared being set exactly level, and heated sufficiently to blue its surface (for the purpose of burning out the tallow with which its interstices are filled, when not in use, to protect it from oxidation), the wooden pattern of the speculum was placed on it, and founders' sand rammed round it to its top. The mould thus formed was about a foot deep, the thickness of the speculum, $5\frac{1}{2}$ inches in this instance, being determined by the quantity of metal melted. By this arrangement, as Dr. Robinson formerly explained to the Academy, the fluid metal which comes in contact with the hoops is *chilled* at once into a dense sheet about half an inch thick ; the air which might be entangled with it in pouring, escaping through their interstices. The circumference sets much more slowly in consequence of the inferior conducting power of the sand ; and the upper surface, which is only in contact with air, remains so long fluid, that the greatest part of the shrinkage occurs there ; its tendency to crack the cast is prevented, and the coarse structure which it produces is confined to a place where it is unimportant.

On this occasion, besides the engrossing importance of the operation, its singular and sublime beauty can never be forgotten by those who were so fortunate as to be present. Above, the sky, crowded with stars and illuminated by a most brilliant moon, seemed to look down auspiciously on their work. Below, the furnaces poured out huge columns of nearly monochromatic yellow flame, and the ignited crucibles during their passage through the air were fountains of red light, producing on the towers of the castle and the foliage of the trees, such accidents of colour and shade as might almost transport fancy to the planets of a contrasted double star. Nor was the perfect order and arrangement of every thing less striking : each possible contingency had been foreseen, each detail carefully rehearsed ; and the workmen executed their orders with a silent and unerring obedience worthy of the calm and provident self-possession in which they were given.

It has been found that a good criterion of the time for pouring the alloy into the mould is afforded by stirring it with a pole of dry wood. This, as long as the temperature is above a certain point, reduces the film of oxide which covers its surface ; and it becomes clean and bright, though a new film forms immediately. At length as it cools this reduction no longer occurs ; and at a signal the three crucibles are emptied into the mould by means of levers connected with the pivots of their cradles. Though familiar with heavy castings, Dr. Robinson had never seen anything so magnificent as the burning lake that was then produced ; and for many minutes it rolled in heavy waves like those of quicksilver, which broke in a surf of fire on the sides of the mould, effecting the most perfect mixture of the metal. At last it became solid, and was examined as it cooled, till it barely yielded to pressure with an iron rod at its centre, which is the indication that it may be removed to the annealing furnace.

This furnace extends along the fourth side of the mould : it is a low square chamber lined with firebrick, with sides about thirty inches thick, strongly hooped, and covered by an arch, from the centre of which rises a flue. Its floor is convex, of the same curvature as the speculum, and is heated from beneath by nine arches, which communicate with lateral flues. It opens towards the mould by a low arch a little wider than the speculum ; but behind has merely an aperture to admit an iron bar. For some weeks the chamber and arches had been kept full of burning turf, so that the whole interior was of a full red heat. The speculum, also red hot (at which temperature, it is to be remarked, the alloy has nothing of that brittleness which characterises it when cold, but is as tough as malleable iron), was cleared from the sand, and encircled by a strong ring attached to the bar above mentioned. By connecting this bar with a powerful capstan, it was drawn from the bed of hoops, along strong beams covered with iron, to its place in the centre of the annealing furnace. The ring was then removed, and the rest of the chamber filled up with charcoal ; the arches with fuel ; all the flues and apertures were closed carefully with masonry, and it was left to cool gradually for sixteen weeks, during the first three of which the exterior of the building was sensibly warm.

In the course of this year considerable progress was made with various parts of the mounting ; and when Dr. Robinson visited it in February, 1843, he found that the speculum had been ground (on a machine similar to the old one, but of strength proportioned to its work) ; that the foundations of the piers were laid, the tube was in preparation, and the massive frame-work and levers by which the speculum is supported in the tube, were cast. This elegant contrivance requires some explanation. Suppose the back of the mirror divided by two concentric circles into three portions, of which the central circle is cut by radial lines into six sectors, the middle zone into nine segments, and the exterior into twelve, and that all of these are equal. If each of these be supported by an equal force applied at its centre of gravity, the speculum is obviously in the most favourable condition as to flexure. The frame mentioned above is rectangular with a cross-piece cast in one, and weighs one ton and a half : it bears three strong triangles, also of cast iron, supported at their centres of gravity on hemispherical bearings. Each angle of each of these bears a similar triangle, the angles of which give the twenty-seven points of equilibrated bearing for the speculum. They do not, however, press directly, but carry platforms of cast iron of the shape of the areas which they are to bear, and made exceedingly stiff by flanches at their edges, and by edge-bars crossing them diagonally. A layer of felt is over these ; strong uprights from the frame of a similar character prevent any lateral shifting ; and the operation of the arrangement is found to be perfect at all altitudes.

The construction of the tubes, the piers, the mechanism of the counterpoises, occupied the remainder of this year and the beginning of 1844. In August, a partial polish was given to the mirror in order to verify its focal length (which was found exactly fifty-four feet); and the observing galleries and the apparatus for controlling the instrument in right ascension were proceeded with. All these gigantic constructions (of whose prodigious mass some idea may be formed from the fact that they contain, along with their other materials, more than one hundred and fifty tons of iron castings), and have been executed in Lord Rosse's workshops, by persons taken from the surrounding peasantry, who, under his teaching and training, have become accomplished workmen, combining with high skill and intelligence the yet more important requisites of steady habits and good conduct. It may also be mentioned that (such was the clear and definite arrangement of the whole in its inventor's mind) nothing failed from first to last ; and it was not necessary for him in any instance to retrace his steps.

At the beginning of February, 1845, the work being sufficiently advanced to permit the use of the instrument without personal danger, Dr. R. and his friend Sir James South were invited to enjoy the trial of it. Its appearance is certainly peculiar, and presents a striking contrast to the more complicated framing of the three-feet telescope which is placed beside it. At first sight, one wonders how it is to be moved, for nothing attracts notice except the massive piers and the tube ; but a nearer approach shows that it is the perfection of mechanical engineering. To have mounted it on the plan of the three-feet would have been impracticable as well as useless. The speculum with its supports is seven times the weight of that in Herschel's four-feet, and both on this account, and the well-known principle that similar machines are weaker in proportion to their bulk, such a stand must have been so heavy as to present great obstacles to its motion. The vast surface exposed to the action of wind must have made it unsteady ; and its durability could not be great. Lord Rosse, therefore, determined to confine the range of observation to the vicinity of the meridian. There the stars are at their greatest altitudes, and atmospheric influences affect them least ; their places can be determined with most accuracy, and an equatorial movement, so essential to micrometer measures, can be easily obtained. With such optical power there will never be a scarcity of objects for examination ; and the restriction will only be felt in the case of planetary bodies. The base of the actual mounting is a very massive joint of cast iron ; its lower axis permitting motion in the meridian plane, its upper in a direction perpendicular to that circle. On this is firmly bolted a cubical wooden chamber, about eight feet wide, in which the speculum is placed, one of its sides opening for the purpose. This again carries the tube, which when vertical and viewed from the interior of the chamber, is more like one of the old round towers than any more ordinary object of comparison. It is fifty feet long, eight feet in diameter in the middle, but tapering to seven at the extremities: it is made of deal staves an inch thick, hooped with strong iron clamp-rings, and secured from collapse by iron diaphragms ; and carrying at its upper extremity the apparatus of the Newtonian small mirror, which, from its great weight and bulk requires to be counterpoised. The telescope is moved in declination by a strong chain cable attached to its top and passing over a pulley fixed at a proper height to the north, down to a windlass on the ground which is wrought by two workmen. East and west, near the top of the piers, large iron pulleys are fixed, having free movement in azimuth, so that their planes may always be in those of the traction : chains suspending the counterpoise weights pass over these to the sides of the tube. The weights, however, are constrained to descend in quadrants of circles by chain guys attached to the frame which bears the declination pulley. It is easily seen that their action is a maximum when the tube is level, and nothing when it is vertical; but between these positions it decreases more slowly than the downward tendency of the tube. To correct this, the tube is connected with loaded levers (placed to its south), by chains of such lengths that one of them is not raised till it is at 40° altitude, and the other at 80°; the latter being necessary for the return of it after it has passed the zenith. The slow motion in declination was not yet applied, but the ordinary one was quite convenient, except for the difficulty of giving orders to the men, who were sometimes seventy or eighty yards from the observer. A two-feet circle, with a fine level and a pair of verniers, will also be attached to the tube to give the declination ; its place was then supplied by a small protractor, five inches diameter, over which was a strong screen to protect the assistant who attended it from any such casualty as the fall of an eye-piece.

The eastern pier bears what may be called the meridian of the instrument : it is a strong semi-circle of cast iron, about eighty-five feet diameter, and composed of several pieces accurately planed. Each of these is bolted to the pier and separately adjustable to a meridian line formed by straining a fine wire over notches in two cast-iron chairs firmly secured at the north and south of the masonry. Sir James South took charge of this delicate operation, and performed it with such precision that when a transit instrument was adjusted by this line, it gave the passage of Polaris *to a small fraction of a second*. The telescope is compelled to move in the meridian, being connected with this circle by a strong bar provided with friction rollers, that it may traverse it easily ; and thus it can be used as a transit instrument with considerable precision. But this bar is racked, and attached to the tube by wheelwork, so that a handle near the eye-piece enables the observer to move it on either side of the meridian, and thus examine it before its passage, or follow its motion. The movement is surprisingly easy ; and a rough graduation on the bar supplies at present the place of an hour circle for finding objects, for which it is quite sufficient, except that the strong light required to set it disturbs the repose of the eye. The elder Herschel has not in the least exaggerated the importance of this when faint objects, especially nebulæ, are to be examined ; and a better contrivance is to be applied. The rack being perpendicular to the meridan, gives a motion not strictly equatorial, but easily made so: had the declination pulley been in a parallel to the earth's axis, passing through the great joint, and had this latter been itself equatorial, this would have been the case ; but the deviation is easily corrected by the addition of a second pulley altering the direction of the chain. Its range is half an hour on each side of the meridian for a star at the equator ; and Lord Rosse intends to effect it by clock-work, as is now generally the case in large equatorials ; though the problem is much more difficult than in those instruments.

The western pier supports the stairs and galleries destined to the observers. Up to 42° of altitude is commanded by the first of them : a strong and light prismatic framing slides between two ladders attached to the southern faces of the piers : it is counterpoised and is raised to any required position by a windlass ; its upper plane affords support for a railway on which the observing gallery moves about twenty-four feet east and west, two of its wheels being turned by a winch near the observer. Three other galleries in succession reach to 5° below the pole ; these are each carried by two beams which run between pairs of grooved wheels, and are drawn forward, when they are turned, by a mechanism of singular elegance. These are able to hold twelve people, but one man can easily work them; and though it is rather startling to a person who finds himself suspended over a chasm sixty feet deep, without more than a speculative acquaintance with the properties of trussed beams, all is perfectly safe. Every bearing part has been *proved* to ten times its utmost probable load, and the doors of the galleries open inward, and are kept close by springs. From this point too is obtained the most distinct perception of the telephone's prodigious bulk, which at a greater distance is not so striking, for want of a standard of comparison ; yet, notwithstanding the hugeness of the masses to be moved, so effective are all the arrangements that Sir James South found it was possible to uncover the mirrors and find a given star in less than eight minutes.

Unfortunately the whole month of February was of the worst astronomical character; and though the great speculum had only the imperfect polish already noted, it was kept in the tube as long as there were any hopes of seeing the great nebula of Orion. That, however, was always clouded while within its range. A few clear minutes on the 13th allowed them to see some stars and clusters ; but the only circumstances worth mentioning were, that it showed the stars of Castor far apart *without an eye-glass*,[1] and that the stars of the cluster 67 Messier, which Sir J. Herschel describes as being from the eleventh to the twelfth magnitude, were many of them as bright as those of the first appear in a three and a half feet achromatic.

At length, when all hopes of Orion were lost in the twilight, the mirror was removed from the telescope, and polished on March 3rd. Its frame is supported in the cubic chamber of the tube by three strong screws which give the adjustment of its optic axis. By unscrewing these, when the tube is vertical, four wheels with

[1] Only twenty-two inches of the mirror could act in this case.

which the frame is furnished come to bear on a railroad fixed at the bottom of the chamber and communicating over a bridge (laid from its door when necessary) with another railway laid on an inclined plane of about sixty feet. It is drawn up this by the declination windlass, and at its top runs on a strong truck by means of which it is drawn a quarter of a mile, on a common road, to the laboratory.

The polishing machine differs in nothing but size from that described by Lord Rosse in the Philosophical Transactions. It makes the speculum revolve once for twenty-four and a half strokes and the eccentric once for eighteen. From seven to eight strokes of twenty-four inches are made in the minute, and the lateral movement of the speculum by the eccentric is fourteen inches.[1] A screw whose nut runs on a railroad above the machine lifts the speculum, with its frame and levers, from the truck, and deposits it on the revolving platform, where it is levelled, centred, and secured. The same apparatus serves to move the polisher during its preparation and to apply it to the speculum, so that it is even more manageable than that of the three-feet was. It was cast with the transverse grooves ; the circular were cut in the lathe. The time required for polishing is about six hours; and Lord Rosse has found that this period cannot be exceeded without injury to the figure, in consequence of the soft pitch being squeezed out, and the harder and unyielding material coming into contact with the iron of the polisher : unfortunately this occurred to some extent in the present instance. The ammoniacal solution of soap used towards the close of the process, happened to be made with ammonia prepared from gas liquor and containing some substance which acted on the mirror and produced a dulness that was not removed till after three hours' additional work. Lord Rosse warned them that the figure must be imperfect, and wished to repolish; but they overruled this proposal, and it was replaced in the tube next day.

On examining it by diaphragm and discs, it was found that, as he anticipated, the edge was not quite perfect. All its zones showed ε Ursæ Majoris very well with 560 ; but the exterior six inches manifested, when the star was thrown out of focus, that though of the same focal length, this portion was irregular. Few other double stars were observed, as most of the lucid interval from the 4th to the 13th of March was devoted to nebulæ, and after that it again became cloudy ; but enough were seen to satisfy them that the instrument possessed a very high defining power. This, indeed, was evident from the admirable exhibition of Regulus, seen on March 5th, neat and round, without appendages or flare. Gamma Leonis, ε Virginis, 2 Comæ, and Gamma Virginis, were also well shown with powers of 400 to 800 on an unfavourable night ; and the companions of η Ursæ, and 245 of Struvé's second Catalogue, which appear in the Slough and Pulkova telescopes as of the eleventh and tenth magnitudes, seem in this large stars.

[1] These are the proportions which Lord Rosse prefers ; but it must be kept in mind that they change with circumstances. *Probably* they will not answer for those specula which have an aperture larger than one-ninth of their focal length, and *certainly* not for those which are perforated in the middle. Dr. Robinson has made many experiments on one of the latter, fifteen inches aperture and nine feet focus, with a machine nearly the same as Lord Rosse's ; and he finds that the nature of the polishing depends on the figure given in grinding. If the eccentric be regulated so as to make this hyperbolic, its action must be lessened in polishing so as to shorten the focus. In this way it is possible to obtain very good results. He, however, prefers the opposite course pointed out by Lord Rosse ; grinding to an elliptic figure, he polishes with a very long primary stroke, and small action of the eccentric. A speculum thus polished shows ε Arietis well separated and defined with 940, and with 465 the fifth star in the Trapezium of Orion's nebula is visible even when the acting surface is reduced to seventy-two circular inches.

Of planetary bodies, none were visible except D'Arrest's Comet and the Moon. The former, when viewed March 10th, presented nothing remarkable: the brighter portion, towards the centre, showed no abrupt change of light which might indicate a solid nucleus ; there was no resolvable appearance in the Coma, and the very minute stars with which that part of the sky was dotted, were visible almost to its very centre. Only one view of the moon was obtained, March 20th, and it was shared with them by several visitors, who, when once in possession of the telescope, were by no means disposed to make way for the astronomers. The fascination of the sight is, indeed, such, that one can scarcely withdraw the eye : Dr. Robinson, therefore, and his friend, had but little time for observation. He was, however, much interested by the vicinity of the craters named Hansteen and Mairan, in the map of Beer and Mædler, where, besides the crowd of hills described by them, there are an infinity of others not visible even in the three-feet, but looking in this with 560 like grains of sand. Are these fragments ejected from the crater ? If so, and if they occur round others, it would explain what had always presented to him a great difficulty. The lunar craters differ widely from those of earth ; and most in this, that their depression below the general surface is enormously greater than the elevation of their walls above it, while the area of the hollow is far greater than that of the latter. What, then, became of the materials which had once filled it ? He had formerly supposed that they were in a fluid or gaseous state when ejected ; but the fact just mentioned seems to give the true solution, and appears to account for them when combined with the consideration of the feeble gravity of the moon, which would permit the exploded fragments to be scattered over a far larger space than with us. Another beautiful object was the river-like valley that runs north-ward from the crater Herodotus : its raised banks, and their irregularities, were easily seen ; the internal and external shadows could have been satisfactorily measured had a micrometer been applied. As it was, the much greater breadth of the former showed at a glance that this strange channel was sunk deep below the lunar surface. Taking as a standard the measures given there by Beer and Mædler, he had no doubt that they then saw *without difficulty* spaces of eighty or ninety yards. It is difficult to say *à priori* what should be the minimum visible at the moon in such a telescope. If we assume, as one extreme, the statement of Amici, that the non-coincidence of two black lines on paper can be seen at twenty-eight feet, when it amounts to one twelfth of an inch, or subtends $51''$, then 311 feet should be visible at the moon with 1000. On the other hand, Jurin states (Smith's Optics) that a piece of silver wire can be seen on white paper, when it subtends $3''$, a result depending on the intensity of this metallic reflection. This would give eighteen feet ! Dr. Robinson finds that he can see the spider lines of his circle without much contrast of light, when they subtend to him $16''$ This gives ninety-seven feet ; but it must be remembered that aperture influences visibility as well as magnifying power, though we cannot as yet estimate its effect numerically.

The most important part of their observations were made on nebulæ ; and, besides establishing completely the prodigious superiority of this instrument over all yet constructed, they have added some facts to our knowledge of these mysterious objects. A list of them was formed from the invaluable catalogue of Sir John Herschel (Phil. Trans. 1833), comprising such as, from brightness or any other peculiarity, seemed derserving of notice ; of which forty were examined by Dr. Robinson and also by Sir James South, except some which the latter lost while making the transit observations required for the meridian line.[1] They may be separated into three classes ; those which are round and of nearly uniform brightness ;[2] those which are round, but appear to have one or more nuclei ;[3] and those which are extended in one direction, sometimes so much as to become long stripes or rays.[4] Of the first class, all that were examined are easily resolved, even with a triple eye-piece of wide field and power 360, used for finding the objects. In 854 the stars were seen through haze ; in 1929 during twilight ; and 1833 was noted as "consisting of rather coarse stars, and resembling Messier 13." Any increase of brightness towards the centre seemed to proceed from the greater depth of stars there rather than from any notable difference of their magnitude. But the second class presents much more interesting phenomena : the appearances which previous observers had described as *sudden condensation, nuclei,* or even single or multiple *central stars,* proving to be clusters of comparatively bright stars, surrounded by much larger collections of minute ones. A very beautiful example of this is 1456, fig. 41, M. 94, described in the catalogue as "very suddenly much brighter, almost up to a nipple-shaped nucleus :" it proved, however, to be "a vast circular cluster of stars, with ragged filaments, in which, and apparently central, is a globular group of much larger stars, power 400." The same system of arrangement (which seems very common) occurs also in 706, 748, 805, and many others : it is also found in the magnificent clusters 1663, M. 3 ; 1558, M. 53 ; and 1916, M. 5. In these, the splendour of which is not described, besides the stars visible in other instruments (which here seem of the first or second magnitude), the whole field is crowded with others much smaller, to such a degree that, had the first been absent, these would still have been noted as remarkable objects. The interior group is not, however, always central or symmetrical, but has knots of greater condensation, which sometimes (as in 1385) are alone visible in smaller telescopes, and then look like "twin nebulæ ;" at others (as in 739), like stars. In 1622, fig. 25, M. 51, which is so well known from a sort of resemblance to Saturn, and from the more exact analogy which, as Sir John Herschel has well remarked, it bears to the Milky Way, we have another different development of this arrangement. Here also the central nebula is a globe of large stars ; as indeed had been previously discovered with the three feet telescope : but it is also seen with 560 that the exterior stars, instead of being uniformly distributed as in the preceding instances, are condensed into a ring, although many are also spread over its interior. Were the centre absent, we should have a ring nebula ;[5] and were the line of vision near the plane of this ring it would

[1] Sir James South published an interesting and instructive notice of this telescope in the *Times*, April 16, 1845.
[2] Nos. 538, 739, 777, 844, 845, 854, 1797, 1833, 1907, 1929.
[3] Nos. 564, 706, 711, 743, 748, 749, 805, 843, 846, 1146, 1385, 1456, 1622, 1881.
[4] Nos. 536, 604, 668, 791, 792, 810, 859, 1066, 1132, 1148, 1352, 1357, 1368, 1466, 1926. The numbers and the figures cited in the text are those of Sir John Herschel's catalogue.
[5] It is possible that the exterior part of M 94, may be merely a circular *disc* of stars : the absence of the central globe would make this a planetary nebula : but it is possible that these differ from the annular only in degree ; all the latter which he has seen having faint nebulosity within them.

become one of those rays with a bright nucleus and parallel band or satellite nebulæ which occur so frequently in the catalogue. In comparing it with our own sidereal system, Dr. R. thinks we should consider the stars visible to the naked eye, and the larger telescopic classes as constituting the central cluster, while the Milky Way represents the exterior and minuter stars either disposed in an irregular ring or in a stratum, two of whose dimensions are much greater than the third. We have no reason for believing that the comparative brightness of stars depends *only* on their distances ; 61 Cygni is not more remote than *a* Lyræ ; much less can we assume that our stars are uniformly distributed : Orion, the Pleiades, Prœsepe, the clusters in Perseus, M. 36 and 37, with many others, are evidently mere knots of condensation in our immediate neighbourhood, our peculiar cluster ; and it seems a mere arbitrary assumption to fancy that, were we transported to a remote part of the Milky Way, we should see anything similar to our present sky.

The nebulæ of the third class which were examined seemed to differ from this type only by being seen obliquely, and therefore projected into ellipses sometimes almost linear. In this last case they proved much more difficult of resolution, probably from greater *optical* condensation, and yielded most easily towards their minor axes. In these the nucleus of brighter stars is sometimes extended like the exterior portion, as in 602, which is of considerable length and easily resolved : the central part has three knots, of which two are represented in fig. 70, all the rest having been invisible. 668 is similar, but the central part is of more uniform character. In general, however, the nucleus is globular, and remarkable from the comparative smallness of its diameter, and its very condensed appearance. Either the stars which compose it are few in number, or more closely compacted than is usual. 1132, M. 98, is a good example : "the long ray is resolved, except at the very extremities, with 560 ; the globular nucleus is seen with 1280 to consist of very close stars." 1148, described as "a nucleus with two branches, a star north following," appeared to Dr. R. as "as irregular ring of stars round a brighter group, but having an appendage like that of M. 51, in which is the bright star seen by H." 1357, fig. 37, is a similar object, both "the ray and appendage being full of stars, but the nucleus requires a higher power to resolve it than the night will bear." In 1466, fig. 84, *the nucleus projects on each side of the ray*, so that its diameter must be greater than the thickness of the exterior stratum.

He could not leave this part of his subject without calling attention to the fact that no REAL nebula seemed to exist among so many of these objects chosen without any bias : all appeared to be clusters of stars, and every additional one which shall be resolved will be an additional argument against the existence of any such. There must always be a very great number of clusters, which from mere distance will be irresolvable in any instrument ; and if it prove to be the case that *all* the brighter nebulæ yield to this telescope, it appears unphilosophical not to make universal Sir J. Herschel's proposition, that "a nebula, at least in the generality of cases, is nothing more than a cluster of discrete stars."

These observations will suffice to show how much may be hoped from this telescope ; but they are far from being a fair measure of its powers, being made at very low temperatures. Almost always the thermometer was at 22° or 20ᶜ when they ceased working ; and on one occasion it was as low as 17°, the lowest he remembered in Ireland. In such circumstances it is notorious that even small reflectors act very imperfectly : and he was therefore unprepared for any tolerable action of this gigantic speculum. In the day time it was of course colder than the air, and, if uncovered before that had sunk to its temperature, was covered with dew : when this went off it always defined sharply. The huge mass of metal cooled much more slowly than the atmosphere ; and as the difference increased, the performance of the telescope was deteriorated. This arose from no change of figure, as he satisfied himself by throwing the stars out of focus ; it was probably the result of currents in the tube occasioned by this difference of temperature. How far it will be possible to obviate this by mechanical means, remains to be tried ; but it is certain that the inconvenience does not increase in a higher ratio than the power of the telescope, as he had formerly apprehended. On the same nights, it defined quite as well as the three-feet with a far lower power; and therefore, it is reasonable to expect that it can be used with advantage much more frequently than he once supposed.

Enormous as is its illuminating power, it might be increased one-third, by using it with the front view, supposing it can be properly figured for this oblique action. Without that, he fears that in an instrument where the aperture is so large compared with the focal length, the definition would be imperfect. He verified this by an experiment with the three feet, and found that though the light was increased quite as much as he expected, yet the perfection of the image was utterly destroyed for large stars. There was no exact focus, but merely two places where the sections of the cone of rays were smallest. One, the least exceptionable, showed a flare in the direction of the slope like a comet's tail : at the other this disappeared, but the star became a sort of curved rectangle with rays from its corners. In the Newtonian form this speculum a few nights before had defined ζ Orionis very well with 500 ; but now γ Leonis could not be seen double with any power ; the companion of Rigel (some way from the meridian however) was lost in the flare, and even that of Polaris, though perfectly visible, was sadly disfigured by it. It was of course useless to try more difficult tests, as even this degree of imperfection would make it utterly incapable of resolving such objects as the nuclei of the long nebulæ, be its illuminating power what it may. One thing, however, deserves notice, that in consequence of removing the second reflection, the colours of the stars come out with extraordinary splendour.[1] β Cygni, for instance, had a pureness and brilliancy of yellow, in the large star, which was new to him, though he had seen it in many first-rate telescopes. Lord Rosse does not apprehend any insurmountable difficulty in applying his method to give the form neccessary for aplanatic oblique reflection : more than one plan for this has occurred to him ; and Dr. R. believes it is his purpose, as soon as the six-feet has its machinery completed, to try them on one of the three-feet specula, and, if successful, to alter the great one.

---

[1] The lenses of achromatics have often a tinge of green or straw colour which modify the colour of objects seen through them. Something of this may perhaps cause the predominance of green and "cinereus" which exists in the Dorpat catalogue.

As it is, Dr. R. congratulates the Academy and their country on the success of this matchless instrument ; to which, as nothing at all approaching to its power has yet existed, so it is not probable that there will soon be any superior. It has been reported that the French Government, at the suggestion of M. Arago, are about to construct an achromatic of a metre aperture. Supposing homogeneous discs of glass can be obtained and wrought of that magnitude, there remain other difficulties. The optician who proposed to supply them stated that they would weigh at least four hundred pounds ; now these, when mounted, must be supported by at most two lateral bearings ; and it is known that a very moderate pressure produces in glass a double refraction, most injurious to its performance in an object glass. But supposing this and the equally probable change of curvature from the weight of the lenses obviated, still such an achromatic would be far below the six-feet in quantity of light. From Amici's experiment with an object glass of two and a half inches it follows that it equals a Newtonian when their acting surfaces are as six to ten : this would imply in the great one an aperture of fifty-six inches and a focal length of eighty feet. But the absorption certainly increases with the thickness of the medium, though neither the law of this, nor the loss by the reflections at the four surfaces, are accurately known. Mr. Potter found that a good object glass by Dollond of four inches aperture and six feet focus transmitted but 0·66 of the incident rays. This gives the ratio of the equivalent surfaces 0·74, and it will be still greater where the glass is three or four inches thick. It is said that the construction of a reflector still larger than this is contemplated by a northern Sovereign who has already shown himself a most munificent patron of Astronomy. If so, none will rejoice more than Lord Rosse himself. It was not the mean desire of possessing what no other possessed, or seeing what no other had seen, that induced him to bestow so many precious years on this pursuit : had such been his motives, he would have kept to himself his methods, instead of opening his workshops without reserve to all who had the slightest desire of following his steps, and communicating in the most liberal manner the fruits of long and painful experience. His sole object is to extend the domain of astronomical knowledge : and the more common such instruments become, the more perfectly will it be fulfilled.

On Lord Rosse's Telescope.

By Dr. T. R. Robinson[1]

THE Rev. Dr. Robinson gave an account of the present condition of the Earl of Rosse's great telescope, and detailed some observations made with it during a recent visit to Parsonstown.

In 1845 he had laid before the Academy the results obtained by Sir James South and himself, at the first trials of that magnificent instrument. The most remarkable of them had reference to what has been called the Nebular Hypothesis, in which it is supposed that nebulous matter forms suns and planets by its gradual condensation. Above fifty nebulæ, selected from Sir John Herschel's catalogue, without any limitation of choice but their brightness, were *all resolved without exception*. From this he conceives himself authorized to ask, is there any evidence that nebulous matter has real existence ?

The appearances which were supposed to indicate the gradual condensation of this imaginary fluid, namely, an increase of brightness towards the centre (sometimes almost looking like a star surrounded by a faint atmosphere), were shown to be caused by a peculiar construction of the systems in which they had been found. This the telescope demonstrated to consist of a central cluster, mostly globular, of comparatively large stars, surrounded by an exterior mass of much smaller and fainter stars, whose arrangement is often circular and thin like a disc. When seen obliquely, they seem like long oval or pointed rays ; and in this case, from the optical condensation of their component stellar points, the resolution is more difficult, but even here it was invariably effected.

He has often been asked why this instrument had given no further results. They who put the question had but a faint idea of the overwhelming pressure which the last three years exerted here on all who were resolved to discharge the duties which men owe to their country. Lord Rosse is not a person to seek knowledge or enjoyment in the heavens, when he ought to be employed on earth ; and he devoted all his energy to relieve the present misery and provide for the future. During this interval some parts of the machinery which could be finished by his workmen without his superintendence, were completed ; a duplicate speculum, which had been previously cast, was ground and polished by them ; but nothing of note was performed except the discovery of the spiral arrangement in 51 Messier, and the resolution of the great nebula of Orion, both of which have been published by Dr. Nichol.

These days of evil are past ; and though the future is still dark and threatening, yet he trusted it would bring nothing but what wisdom and benevolence might

[1] Proceedings R. I. A. IV, 119 (March 1848).

turn to good ; and in this same hope Lord Rosse felt himself at liberty to resume his favourite pursuits. Dr. Robinson found the new speculum imperfectly polished, and the old one tarnished by wet, which had found access to it while it was not attended to. No difficulty was apprehended in repolishing ; but for a long time the process failed unaccountably. The figure was hyperbolic, and the surface irregular. This last can be easily ascertained during the operation. For the first two hours, the peroxide of iron used as the polishing material covers the surface with scratches, which gradually disappear afterwards. If these be examined by the reflecting image of a lamp or window, when the work proceeds well they appear as dark lines, otherwise they show a luminous edge indicating a curvature of the adjacent surface ; and whenever this occurs, the definition will prove imperfect. Weary at last of these trials (each of which involved four days of hard work), Lord Rosse determined to experiment on one of the three-feet specula, which, as Dr. Robinson formerly explained to the Academy, could be examined on the engine, by a dial placed above the tower where it stands.[1] Here also there were five or six failures, till Lord Rosse noticed that there was a difficulty in keeping the speculum properly coated with peroxide of iron ; and the disturbing cause was soon detected. The pitch of which the polisher is made possesses the requisite consistence only at the temperature of 55°. At that time, however, it was below freezing, and it was necessary to warm the laboratory by stoves. The air of that room, therefore, became drier, and evaporated the moisture from the speculum and polisher too rapidly. On examining this with the wet-bulb hygrometer, they found in one instance 17^{u} difference. This was remedied by a jet of steam from a small pipe connected with the boiler of a steam engine, which was regulated so as to keep the air nearly saturrated with moisture ; and *at once* all difficulty was removed. The speculum defined the dial-mark quite sharply with a power of 3800, and, when placed in its tube, left nothing to be desired. The six feet was polished with equal success next day, February 16.

Originally the movement in right ascension was given through a handle moved by the observer. This was found inconvenient ; and the apparatus is connected with a drum below, moved by a workman. It is found that this will afford a ready means of mechanical movement by clock-work, which is now in hands. The arrangement chosen by Lord Rosse is a gigantic *metronome*, the pendulum of which will carry a graduation or polar distance, to which the assistant will set the sliding weight at the same time that the telescope is set on an object. It has been ascertained by trial that the elasticity of the impelling band (100 feet long) is quite sufficient to equalise the intermitting movement of such a scapement.

---

[1] If a lucid point be at the distance d from a parabolic mirror, whose focal length $=f$, its image is formed at a distance from the principal focus for central rays,

$$z = \frac{f^2}{d-f} ;$$

for a zone whose distance from the axis $=y$: this distance is further increased by

$$\zeta = \frac{\tfrac{1}{2}y^2}{d-f}\left\{1 + \frac{y^2}{8f^2}\right\} :$$

ζ can be measured for different zones, *and if it have this value*, the speculum must be parabolic.

In searching for known objects, there is, of course, occasional difficulty in finding them, from the small field of view of ordinary eye-glasses. This is remedied by a supplemental eye-piece of very wide field ; a slide carries it, and the holder of the others, so that by a little shift one can be substituted for the other in an instant. The eye-piece is similar to one which Dr. Robinson had long since made for his own instrument. It consists of three lenses; and fulfils the three conditions of equal flexure of the pencils, achromatism, and flat field, while its distinctness is equal to a Huygenian of equal power. In this one the field-glass is six inches diameter ; it magnifies 216 times with a field of *thirty-one* minutes; and though this will only bring into action forty-three inches of the mirror, yet even so its optical power is very great; and Dr. Robinson thinks the view of the moon in it the most magnificent spectacle he ever saw. A nebula is easily found in this wide field ; and bringing it into the centre, the eye-slide is shifted, and it is viewed with the higher powers.

The micrometer also appears deserving of notice. Notwithstanding the prodigious light of the telescope, it was found that any illumination of wires extinguishes many of the fainter details in nebulæ. Lord Rosse, in drawing 51 Messier, used a very ingenious substitute, a screw whose threads were rubbed with phosphorus. Dr. Robinson had made experiments with a micrometer whose platina wires were faintly ignited by a voltaic current ; he found, however, that the heated air produced tremors quite incompatible with the use of high magnifying powers. An experiment of Mr. Babinet, described in the Comptes Rendus, has suggested a contrivance which seems quite satisfactory. If light be admitted through the edge of a piece of parallel and pellucid glass, *it cannot escape through its faces*, because it is incident on them at an incidence greater than that of total reflection. Looking through the faces, therefore, the field of view is absolutely black, unless there be bubbles or striæ in the glass ; but if a scale of any kind be engraved on either of them with a diamond, the light escapes through the cuts, and they appear luminous. The division are 6″, and the eye-piece has, of course, a position circle.

Dr. Robinson regrets that he had very few opportunities, while at Parsonstown, of using the telescope, in consequence of the unfavourable weather, and of the circumstances which have been stated. Most of them, too, occurred while the speculum was imperfect ; yet some facts which he observed may be worth the notice of the Academy.

In the moon may be mentioned that the wide surface at the bottom of the Crater Albategnius is all strewed with minute blocks, not visible in the three feet with 500. The exterior of the mountain Aristillus is all hatched as it were with deep gullies radiating towards its centre ; and he was able to confirm his former observations, that the bright streaks which radiate from some craters (Kepler in this case) are not raised above the surface.

Jupiter was several times seen. The dark brown belts presented, on February 20, a remarkable appearance ; they were full of faint striæ running nearly parallel to them, and seemingly belonging to the brighter zones on each side. The colour of the belts is deepest at the centre, and gradually dies away towards the edge. This he regards as evidence that they are seen through an atmosphere of considerable depth and imperfect transparency. From this too, and from the fact that the polar regions present a similar though less intense shade, it is evident that the darker parts are the body of the planet, and the brighter its clouds.

Several nebulæ, in addition to those which were mentioned in Dr. Robinson's former communication, were examined. Of these, Nos. 505, 540, 668, and 988 of Herschel's catalogue are mere globular clusters; 65 and 66 Messier are of the other class, which he considers to be central clusters, surrounded by discs of smaller stars seen obliquely. The first, however, is less elliptic than in Herschel's fig. 53. 1 Messier was examined, but little addition can be made to Lord Rosse's description of its appearance in the three feet,[1] except that the "nebulosity" is all resolved, and "the resolvable filaments" consist of pretty large stars. There is, however, in the body of the cluster one so much larger than the rest, that it can hardly belong to their system.

The great nebula of Orion was completely resolved in those places which presented the mottled appearance, even in indifferent nights, and while the speculum was imperfect. On February 20, after it was in good order, a power of 470 showed the stars quite distinct there on a resolvable ground; and this clearly separated into smaller stars with 830, which the instrument bore with complete distinctness. The diffusion of so many knots of stars through a vast stratum of others much more minute is a most wonderful sight; and while looking at it he could not help speculating on the aspect which the heavens would present to an observer there. Yet, possibly, the Milky Way, if viewed from without, in the direction of Taurus, would exhibit something similar. The Magellanic Clouds, as described by Sir J. Herschel, are evidently analogous systems. On the same evening an eighth star was found in the trapezium, a seventh having been discovered on the 10th; the first near Herschel's a, and in the opposite direction from the sixth one detected by Sir James South's large achromatic, and more distant; the second near β. It is worth mentioning, as illustrative of the effect of previous knowledge on vision, that having ascertained the parts where the stars were most distinct, he was able to see them in the three feet with certainty; though in former years he had repeatedly scrutinised it for this very purpose in vain.[2]

Two remarkable exceptions to the general plan of nebular systems are afforded by 64 Messier, and h 464. In general the centre is occupied by a cluster of comparatively large stars, round which the others are grouped. But in the first of these (Herschel's fig. 27) there is a central vacancy looking absolutely black by contrast with the surrounding mass of stars. At its south and preceding *edge* are disposed, rather irregularly, a knot of about 100 larger stars, of which it is scarcely possible

[1] Phil. Trans. 1844, p. 322.

[2] A recent notice mentions that Mr. Bond, of Harvard University, in the United States, has resolved parts of this nebula with a Munich achromatic similar to that of Pulkova. The climate and lower latitude would assist him in some degree; but Dr. Robinson thinks this success must be in a great measure due to that precise knowledge of the phenomenon, and of the points where it might be looked for, which is afforded by Dr. Nichol's work. He perceived the fifth and sixth stars of the trapezium, but saw nothing of the new pair. It must be remembered that, however sharply an achromatic may define objects whose light is intense, its illuminating power is far inferior to that of a large reflector. An object-glass of sixteen inches has not as much light as a *Newtonian* of twenty-one.

D

to doubt *that they had once formed the usual globular cluster in the vacancy*, and had been in some way displaced from it. The second is a fine planetary nebula *in* the splendid cluster 46 Messier. The stars of the latter are large and very brilliant, so that probably it is not very remote ; but the other is a round disc, entirely composed of minute blue stars, without any condensation in the middle ; and the singularity is, that it is not encroached on by the stars of 46 Messier. One very large one is near its edge ; but evidently it would not be possible to describe a circle of equal diameter in any other part without including several. Are we to suppose that this is a case of mere optical connexion? The probability is very small, of a cluster such as 46 Messier (which is not common), and a large planetary nebula (which is very rare) coinciding ; and if we combine with this the probability of a round cavity through one being exactly the size of, and in a line with the other, that probability will be evanescent. It seems, therefore, necessary to conclude, that both are parts of the same system, and possibly more examples of the kind may be found.

Two other clusters, 37 and 50 of Messier, besides their own marvellous beauty, interested Dr. Robinson on another ground ; they are in the Milky Way, and, therefore, are seen on its stars, and at a place where its depth is nearly a maximum. Now, these stars were all of notable size and brightness, so that the telescope evidently penetrated far beyond their enter or limit. This seems to require a change in some of the reasonings in Struvè's admirable Etudes d'Astronomie Stellaire. The author, among other curious matter, by applying the theory of probabilities to the numbers of stars of each magnitude in Argelander's Catalogue, and Sir W. Herschel's Star gauges, and by assuming that all stars are nearly equal, and that the Milky Way is unfathomable by telescopes in its greatest extension, finds this result, that the distance of the sixth magnitude is about seven times that of the first, and that the smallest stars visible in the eighteen-inch reflector of Herschel are $25\frac{1}{2}$ times as remote as the sixth magnitude. But this telescope should show stars at three times that distance, and hence he infers that the "heavenly space" is not perfectly transparent. It appears to Dr. Robinson that the last of these assumptions is inconsistent with the above-mentioned observation ; and that the other is equally at variance with the arrangement so often referred to, in which the central stars are much larger than the exterior. It may also be added, that the *penetrating* power of a telescope does not depend on its light alone, for every one knows that a high magnifying power shows small stars, which are invisible in the same telescope with a lower one. The "sweeping power" was only 157, and though it was the best for finding nebulæ, it was much too low to give the utmost range of vision.

But far the most singular objects which he has seen are the nebulæ which exhibit a spiral arrangement. He re-examined 51 Messier, Herschel's fig. 25, in which Lord Rosse had first seen it, and fully verified it: he could not, however, satisfy himself that it was to be traced in the three feet. On the night of March 11 (the only fine one, by the way, which occurred during his stay), he found several others, of which, however, it is difficult to give an idea without drawings.[1] In 99 Messier the centre

---

[1] Drawings of M 51, M 99, M 97, and h 731, were exhibited.

is a globular cluster, surrounded by spirals, in the brighter parts of which stars are seen with 470 : these have the same direction as in Messier 51, namely, from *east to west*, in receding from the centre. But these are combined with traces of another system in a reverse direction. h 604 is also spiral, but without any other peculiarity. 97 Messier is a strange object. With the finding eye-piece it looks like a figure of 8 with a star at the intersection, but with 470 it is spiral with two centres. The principal one still looks like a star, but with 830 gives a suspicion that it is a very small cluster.[1] The spirals related to this have the same direction as the former; but the other centre, which also looks like a minute star, has a smaller set in the opposite direction. Lastly, h 731, his fig. 43, in which the stars seem larger than the preceding, but in which no central cluster was observed, has curved dark bands across it, looking so like the section of a turbinated shell as to induce a suspicion that this has a similar arrangement, but is seen edgewise.

On the dynamical condition of such systems it would at present be idle to speculate ; it must evidently be much more complicated than that of the ordinary globular clusters, which themselves are intricate enough. Their resemblance to bodies floating on a whirlpool is, of course, likely to set imagination at work, though the conditions of such a state are impossible there. A still more tempting hypothesis might rise from considering orbital motion in a resisting medium ; but all such guesses are but blind. He believes it is Lord Russe's purpose to make drawings of all these, *based on rigorous measurement*, which may serve as evidence of change hereafter, should such occur to any perceivable extent during the ages that are yet to come. The instrument will henceforward by regularly employed by an assistant, whom Dr. Robinson has trained for the task, and on whose zeal and steadiness he can rely ; and as it cannot be turned to the sky without revealing something wonderful and glorious, he is certain that it will yield an unfailing treasure to science, that it will realise the high hopes of its generous master, and be one of the proudest distinctions of his country.

Observation of the Nebula, Herschel 44

By Dr. T. R. Robinson[2]

THE Rev. Dr. Robinson next proceeded to notice a fact of some interest which he lately observed with the Rosse telescope. It related to a remarkable planetary nebula, Herschel's figure 44. This looks, in smaller instruments, like an oval disc, reminding one of the planet Jupiter; but it appears to be a combination of the two systems which he had formerly described. In both these the centre consists of a cluster of tolerably large stars : in the first, surrounded by a vast globe of much smaller ones ; in the other by a flat disc of very small stars, which, when

[1] The next power is 1550, but it was impossible to use it effectually without a clock movement. This is also the case with single lenses, which are particularly effective on such objects.
[2] Proceedings R. I. A. IV, 236 (Nov. 1848).

seen edgeways, has the appearance of a ray. Now this nebula, which he had recently observed through Lord Rosse's telescope, has the central cluster, the narrow ray, and the surrounding globe. He would also add, as a remarkable proof of the defining power of this vast instrument, that he saw with it, for the first time, the blue companion of the well-known γ Andromedæ, distinctly, as two neatly separated stars, under a power of 828. It was discovered by the celebrated Struve, with the great Pulkova Refractor, and is a very severe test. He further wished to mention that, as La Place had anticipated, the ring of Saturn, which was quite visible, showed irregularities, which are most probably mountains, on its eastern side.

Contents of an Ancient Bronze Vessel, in the collection of the Earl of Rosse

By Dr. T. R. Robinson[1]

THE Rev. Dr. Robinson then read the following communication descriptive of the contents of an ancient bronze vessel found in the King's County, and now belonging to the collection of the Earl of Rosse. The antiquarian relics contained in this vessel comprised several celts, some spear-heads, gouges, and curiously constructed bells; they were composed of a beautiful hard bronze, in very fine preservation. The composition of the metal itself, and the style of workmanship evinced in the various articles, argued no mean degree of metallurgic skill in their fabrication. Several of those interesting relics were exhibited to the members; and drawings, which were pronounced to be admirable in their fidelity and minuteness, were displayed of the several implements of way and husbandry which were not exhibited.

"Several years ago, I remarked this vessel in the collection of the Earl of Rosse, and the singularity of its contents made me suppose a description of it might interest the Academy. I, however, found it impossible to acquire any information as to the locality or time of its discovery till now. It had been purchased for Lord Rosse, about sixteen years since, by an inhabitant of Parsonstown; but the men who had found it, with that strange suspiciousness that is such a familiar feature of the Irish peasant, had made him promise to keep the details secret during their lives. The last of them died this winter, and then Mr. —— felt himself at liberty to give me this information. It was found in the townland marked Doorosheath in sheet 30 of the Ordnance map of King's County, near Whigsborough, the residence of Mr. Drought, in what appears from the description to have been a piece of cut-out bog, about eighteen inches below the surface. No river is near the spot; no bones or ornaments, or implements of any kind, were near it; though, had any gold or

[1] Proceedings R. I A. IV, 237 (Nov. 1848).

silver been discovered, the finders would probably not have acknowledged it to any one. I could not learn in what position it was found.

"A very good idea of the appearance of this vessel is given by fig. 1 of the accompanying drawing, for which I am indebted to Arthur E. Knox, Esq.[1] The scale is one-third of the original, and he has given very precisely the actual condition of its surface. It is composed of two pieces, neatly connected by rivets. The bronze of which the sheets are formed possesses considerable flexibility, but is harder than our ordinary brass; and it must have required high metallurgic skill to make them so thin and uniform. On the other hand, it is singular that, neither in this nor any other bronze implements with which I am acquainted, are there any traces of the art of soldering: if it might be supposed objectionable in vessels exposed to heat, yet in musical instruments this would not apply.

"Such vessels have often been found, but the contents of this are peculiar. When discovered (without any cover) it seemed full of marl, on removing which it was found to contain an assortment of the instruments which may be supposed most in request among the rude inhabitants of such a country as Ireland must have been at that early epoch. A few were given away, one of each, in particular, to the late Dean of St. Patrick's, and these are probably in our Museum; but the following remain:

"1. Three hunting horns, with lateral embouchure, shown on the scale of one-third at fig. 2 (D. 656).

"2. Ten others of a different kind, fig. 3 (D. 653); these differ considerably in size, but that represented is of the average size. Some of the largest have the seam united by rivets; in others it is marked by a paler line in the bronze, which seems as if they had been brazed, but is probably owing to a thin web of metal, which penetrated between the halves of the mould in which they were cast. All of this kind seem to have had additional joints, of which three were found, figs. 4 and 5 (B. 963); at least, no other use of these pieces occurs to me; and in none of them is there any convenient embouchure.

"4. Thirty-one bells of various sizes, figs. 6 (B. 945) and 7 being the extremes; of the real size. They have loose clappers within, and many of them slits to let the sound escape more freely. The bronze in these is much harder than in the preceding, and has resisted decomposition almost entirely. I think it can scarcely be doubted that these were bells for cows and sheep, which would be specially useful among the dense forests which then overspread the island.

"5. Thirty-one celts, of very different sizes, but none sufficiently large to induce a belief that they were used in war. In many of them the colour of the bronze is such as, at first sight, to excite an opinion that they were gilded. There are

Two of the size of fig. 8 (B. 244).
Seven of the size of fig. 9 (B. 357).
Six of the size of fig. 10 (B. 350).

[1] This vessel is very similar to one in the Museum of the Academy, which is marked D. 551. As several of the other objects described by Dr. Robinson resemble the specimens contained in the Museum, a reference to the latter is given in each case. [The figs. referred to are not reproduced].

Five of a size intermediate between these, and

Six of the size of fig. 11 (B. 270).

Five of the size of fig. 12 (B. 276).

"It is worthy of notice that in all the points are entire and sharp, and the edges unbroken, and not seeming to have been ever used.

"6. Three gouges, fig. 13 (B. 181). These are, I believe, of comparatively rare occurrence, and therefore were, probably, of less extensive use than the celts; just as the common carpenter's gouge is with respect to his chisel, to which I believe the others to have been the analogues. Their round edge is well adapted either for paring or for excavating bowls and goblets.

"But the finest specimens of workmanship are the spears, twenty-nine in number. These also are of various sizes, and of greater diversity of pattern than the other implements. There are

Two of the size of fig. 14 (B. 54).

Four of the size of fig. 15 (B. 38).

One of the size of fig. 16 (B. 35).

Seven of the same size, but a plainer pattern.

Nine of the size of fig. 17 (B. 34).

Six of a size two-thirds the preceding, but which it did not seem necessary to draw.[1]

"These, also, have their points and edges perfect, and seem never to have been used; they show not only that the workmen who made them were perfect masters of the art of casting, but also that they possessed high mechanical perceptions. If these weapons and the bronze swords (of which our Museum contains several) be compared with those used in our army, it will easily be seen that the former are constructed on principles far more scientific. Some of these may not be obvious to the ordinary reader, as they depend on the properties of bronze. This alloy, especially when in the proportion used for weapons (in which it is an atomic compound, containing fourteen equivalents of copper and one of tin or nearly eighty-eight and twelve by weight, and possess a maximum specific gravity considerably surpassing either of its elements), combines great strength and toughness, but has not hardness to take an effective and permanent edge. It has, however, been

[1] It is a curious circumstance that *six* kinds of spear heads should have been found. Dr. R. had met with seven different names for this weapon in Irish; but as his knowledge of this language is very limited, he availed himself of the high authority of Mr. Eugene Curry, who gives them:

Laighin, pronounced Loy-en.
Sleagh, ... Shle.
Manair, ... Mon-eesh.
Cruireach, ... Crusheach.
Fogha, ... Fow-gha.
Gae, ... Gaé.
Gabal, ... Ga-val.

With the remark, however, that Sleagh and Gae are sometimes used indiscriminately. The Laighin was of foreign introduction, and peculiar to the men of Leinster; it was, therefore, not likely to be used in this locality, so that the collection, probably, comprised all that were in demand. Among these names, four are evidently of Hebrew affinity.

The second is identical with שלח (shlech), a missile spear; the third comes from מנת, fate; the fourth, or rather its abbreviate form, Cruich, is from כרת (chreth), to destroy; the sixth is little altered from קין (kain), a dart; and the last, possibly, comes from גוב, to divide. Mr Curry remarks, also, that several of these names are now given to agricultural instruments; the loy and slaine are familiar examples: Manair now means a mason's trowel. It should seem that metallurgy was made the minister of war long before it became subservient to the arts of peace.

shown by D'Arcet, that if its edge be hammered till it begins to crack, and then ground, it acquires a hardness not inferior to the common kinds of steel, and is equally fitted for cutting instruments. Now, in fig. 14, the strong central cone of bronze, remaining in its ordinary state, effectually stiffens the weapon against fracture; while the thin webs on each side have evidently been subjected to this or some similar process, for their edges are much harder, as well as brittle. In the smaller weapon, fig. 17, the web might be too thin, and, therefore, it is reinforced by a pair of secondary ribs; and in fig. 16, the most highly finished of them all, by four such. It is, however, possible that these ribs may have answered another purpose; they have so strong a resemblance to those on some Malay krises, that they may have been designed, as in those weapons, to retain poison. This practice, I fear, was not unknown among the ancient Irish, as, indeed, it seems to have prevailed among all the Celtic and Iberian races; thus, in the poem on the death of Oscar, published by Bishop Young, in the first volume of our Transactions, the spear of Cairbre is expressly said to be poisoned (nime), and nothing seems to require a figurative sense of this epithet to be understood.

"The most obvious hypothesis respecting this curious assemblage of objects is, that they were the property of some individual, who concealed them in the bog, perhaps on the approach of a predatory party, and perished without recovering them. Against this is the fact that the tools and spears seem not to have been ever used, and the probability that, in such times, every spear-head would have been mounted, and in the hand of a combatant. It seems more likely that the collection was the stock of a travelling merchant, who, like the pedlars of modern times, went from house to house, provided with the commodities most in request; and it is easily imagined that, if entangled in a bog with so heavy a load, a man must relinquish it. And this is connected with another question, the source from which the ancient world was supplied with the prodigious quantities of bronze arms and utensils which we know to have existed. This caught my imagination many years since, and I then analysed a great variety of bronzes, with such uniform results, that I supposed this identity of composition was evidence of their all coming from the same manufacturers. Afterwards I found that the peculiar properties of the atomic compound already referred to are sufficiently distinct to make any metallurgist, who was engaged in such a manufacture, select it.[1] It also appeared to me more permanent in the crucible than when of higher or lower standard. But the same conclusion is forced on us from another ground. Bronze contains tin; now this metal, for all commercial purposes, may be said to be confined to the south-west of England,[2] and, therefore, the bronze trade must have originated with persons who were in communication with Britain. But in ancient history we find only one people of whom this can reasonably be supposed, the Phœnicians, who, like ourselves, seem to have been the great manufacturers and merchants in olden time. That they had factories, if not colonies, in Spain, at a very early period, is known

[1] The technical importance of atomic proportions is remarkable. Speculum metal is 4Cu + 1St; gong metal is 8Cu + 1St; that referred to is 14Cu + 1St; the hardest metal used for cannon is 16Cu + 1St.

[2] There are tin mines in Malacca, but we have no evidence that they were worked so early; and if they had been, it is quite improbable that their produce found its way to the Mediterranean.

to all; and it seems most unlikely that such enterprising navigators would stop there. Of course, one can attach little weight to the remote traditions of Irish history, if unsupported by other probabilities ; but the traces of Phœnician intercourse which they exhibit are borne out by the admixture of Punic words in the language, and by usages which show that the worship of the god Baal, and other Sidonian rites, had once prevailed in the island. Their traffic in amber proves that they must have gone yet further, even to the Baltic; for then, we may be sure, the land carriage of precious materials through various and hostile regions was almost impossible. All, too, that we know of early antiquity shows that they had the bronze trade in their hands. Even down to the time of Aristotle, tin was described by the epithet 'Tyrian'; and in every nation where bronze was in common use, their presence can be traced or inferred. In Egypt, where this compound was of universal use, we know that the people were little addicted to maritime pursuits ; while they were in close communication with the Sidonians (of the same race), through the Mitzräite colony of the Philistines. In Etruria, not less remarkable for its profuse employment of bronze, we know that they did not obtain it directly, for it is recorded that an expedition was fitted out by them, to open a communication with the tin islands, which failed, in consequence of the jealousy of the Phœnicians. Hence we may conclude that the latter held a monopoly of the tin. In Judea, we find Solomon obliged to employ a Tyrian founder for the bronze works of the temple, and we gather from the account, also, how they were cast—in loam.[1] But Greece, in the Homeric age, presents a state of things much more conformable to what I suppose was the condition of Ireland when this collection was buried. Iron scarcely appears to be in use; and it may be surmised that the art of working bronze itself was not generally understood, from the poet's description of Vulcan making the arms of Achilles. No mention is made of casting or moulds, though a reference to Milton's splendid description of the infernal palace shows how much more poetic that would have been than the hammer and anvil. It seems as if the god merely heated and chased into shape sheets of metal, already prepared.[2] It may be added, that Homer describes all articles of superior workmanship as Sidonian ; and represents this people as trading in every part of Greece. Their ships run into some cove, and their factors go to the dwellings of the neighbouring chiefs. These, though at feud among themselves, and driving each other's cattle on every opportunity, receive the strangers kindly, and purchase from them hardware, jewellery, articles of dress, and toys, in return for cattle and slaves. Now, just such a person I suppose the possessor of this vessel to have been, and of this very nation. Commerce was probably carried on in this way along the shores of the Mediterranean, till the destruction of Tyre by Nebuchadnezzar destroyed it also for a time, and then removed its most powerful centre of action to Carthage. That state seems to have chiefly directed its attention westward ; and it is a

---

[1] Moulds for celts have been found here and in other countries, but were, perhaps, employed to recast old bronze ; they could not turn out work very neat, and many of these tools have apparently been cast in sand. These spears were, I think, cast in loam.

[2] Bronze is brittle at a red heat ; but it and even bell-metal are malleable at a temperature below visible ignition. Speculum metal is not brittle while red hot.

confirmation of my opinion that the bronze trade was almost exclusively Phœnician, that about this time the use of the alloy rapidly gave way to iron and steel. In fact, the supply being cut off from Greece and Asia by the ruin of the Tyrians, *they* were obliged to seek other resources ; but in Ireland and other Atlantic lands the traffic must have continued, nay, perhaps, even increased, in consequence of that event, till the fall of Carthage finally cut it off. I would also throw out another suggestion, though at considerable risk of being thought a dreamer. We see in Homer that the Phœnician traders were quite ready to have recourse to violence when they could profit by it ; and, from more historic sources, that, in Lybia and Spain, they took an early opportunity of turning their factories into forts, and enslaving the natives. Did the same thing happen *here*, when the Tuatha De Danaan, a tribe rich in metallic ornaments and weapons, subdued the ruder Firbolgs, who referred their superior knowledge to magic ? Were these shadowy personages also Phœnicians ? Their name signifies 'the tribe of the gods of the Dani or Damni.' If the first, it might indicate Odin and his Asæ; but, besides that *they* must have been far later, it seems highly improbable that such fierce warriors would have been overpowered by any Celtic immigration. If the second, the Damni the inhabitants of Devonshire and Cornwall, must have been completely under the influence of the Phœnician agents, and may at first have imagined and called their accomplished visitors deities. In these Ogygian regions we must not reckon dates too closely; but I believe it is held that the battle of Moytura, which established their dominion, was fought about 500 years before our Lord, and, therefore, at the very time when the fall of Tyre may have been supposed to scatter its people, and the ruin of their commerce incline them to desperate adventure. It is possible that this conjecture may be established or disproved by a comparison of the skulls found in the sepulchral monuments on their battle-fields with those of Tyrian or Carthaginian origin, if any such are known to exist."

44

From Reports of the
BRITISH ASSOCIATION FOR THE ADVANCEMENT OF SCIENCE

Presidential Address by the Earl of Rosse[1]

GENTLEMEN,—I am sure no one can feel sensible of the kindness of my noble friend, in condescending to notice my very humble exertions in the cause of astronomical science, and no one more conscious that the compliment so flattering is undeserved, and, I must say, that I should be but too happy were it now in my power to resign into his abler hands those duties which have just devolved upon me ; for in that case I am sure the Association would have nothing to desire. But as that is impossible, and it has been of late the practice of those who have occupied the position in which I find myself most undeservedly placed, to offer a few observations on the subjects of the Association at the first General Meeting, I feel I have no other course but to solicit most earnestly your kind indulgence. Such a request you would not perhaps consider unreasonable for any one who laboured under the embarrassment necessarily arising from the consciousness of his own inability adequately to discharge the duties entrusted him, augmented, as it must be, tenfold, by that awe which it is impossible not to feel in the presence of men the most distinguished in the varied departments of human knowledge. But perhaps, in this instance, your kindness will allow there is an additional claim to your indulgence. This very embarrassing position is not of my own seeking. To have aspired to the high honour of presiding at one of your meetings would have been an act of presumptuous vanity, which I never did, which I never could have contemplated. A communication from Manchester, announcing that the Association had actually made their selection, was the first intimation which reached me that my name had even been thought of. Under such circumstances, to have declined the honour, and to have shrunk from the responsibility, would, in my opinion, have been inconsistent with proper respect : it remained, therefore, but to endeavour to do the utmost, trusting that your kindness would overlook all omissions, and that the vigilance of the many most able men who guide the proceedings of the Association would detect and correct all important errors. But, however arduous the task, however painful the duty of addressing a meeting so constituted as this is, it is impossible not to participate in the gratification which all must feel in seeing

[1] B. A. Report, 1843, xxix.

so many men of eminence assembled to assist each other in promoting objects of such deep and general interest.   The man of the world who, busied in the changing scenes of life, watches with fixed attention the actions of men, while he occasionally perhaps casts a passing glance at science as it happens to present to him some new wonder—he cannot fail to look with surprise, and, I may add, with gratification, at a meeting so large (and in this country too), from which politics are altogether excluded.   Here he will see no angry conflict of passions, none of that feeling of bitterness and animosity, which never fails to attend the contests between man and man, between different classes in the same country, or between different nations : all proceeding from the same cause, or nearly so—a struggle for power; in other words, a struggle for dominion over man, and through him over the material things of this world.   But in such a contest, what is gained on one side must be lost on the other.   Here, on the contrary, however much may be gained, there can be no loss to any one.   This is no paradox ; for here the object of the contest is to increase man's knowledge, and with it at once his power over the material things of this world.   It is plain, therefore, that in the objects we have in view, all have an equal interest ; that the contest we are engaged in is one of friendly rivalry, all competing in their efforts to promote that knowledge, that science, which has been given to us as the reward of industry, and by which the gifts of a bountiful Providence may be increased and improved, for the benefit of man, to an extent almost unlimited.

But, Gentlemen, there are perhaps many here who have not been present at other meetings of the Association, who know nothing of the objects actually accomplished by it, and who are not acquainted with the records of its proceedings annually published.   The question, therefore, may be asked, Does this Association actually promote the advance of science, and if so, by what means ?

For a complete, detailed, and triumphant answer to such a question, I must refer to the printed Reports of the proceedings.   It would be unpardonable on my part to take up your time in endeavouring to perform a task, no doubt imperfectly, which has been achieved in the most complete manner by the very able men who on former occasions have undertaken it.   I shall therefore only mention, that original researches in various departments of science, and on a great scale, have been carried on by the Association, upon which large sums have been expended under the most skilful management, and with very important results.   The sum so expended exceeds 8000*l.*   Much also has been accomplished for science, by the resources of the State applied under the advice of the Association ; and within a few days it has been officially announced that the sum required for an important astronomical object, the publication of the Observations of Lacaille and Lalande, has been granted by the Government

For the previous reduction of the observations we are indebted to the zeal, ability and public spirit of Mr. Baily and Mr. Henderson, two members of the Association, who gave their services gratuitously, and took upon themselves the laborious duty of superintending the work. The actual expense incurred, amounting to 1400*l.*, was defrayed out of the funds of the Association.

I am also happy to be enabled to announce, that with respect to another great undertaking you all have heard of, which has been carried on at the public expense, under the gratuitous superintendence of a distingushed philosopher,[1] a most favourable notice has been published by a foreign geometrician of eminence : that notice, or essay, perhaps I should call it, will appear translated in the next number of the Scientific Memoirs. I regret I have not been able to procure a copy of the original essay, and therefore cannot say anything more precise about it ; still I cannot refrain from mentioning it as a subject of much interest in the scientific world. In addition to the researches carried on by the Association, much has been done to aid research. A very important series of papers has been written and published in the annual volumes, under the head of " Reports on Researches in Science." Each of these reports is, in fact, a complete and accurate general view of the actual state of that science, or branch of science, to which it refers, briefly, but profoundly, touching upon every point of interest, so that the man about to undertake the task of endeavouring to advance any particular branch of science may at once, by referring to one of these Reports, know where to look for that information which is indispensable to success, namely, an exact knowledge of all that has been done by others.

These Reports are so numerous, and embrace so wide a field, that to give any analysis of them within reasonable limits, would be impossible ; and to form an adequate estimate of their importance, it is absolutely necessary to examine them in detail, just as they have been published. However, it appears to me, that without presupposing any knowledge whatever of these matters, or of the past history of this Society—without assuming that it has in any one instance effected, by joint co-operation, important and laborious researches in the cause of science, still that, even to a person who will not take the trouble of inquiring and informing himself, an answer to the question, Does the Association advance science ? may be returned, short but conclusive. The answer I should give would be this : I appeal to the experience of every man at all conversant with the history of science, and with the working of scientific societies, whether it is not indisputable fact, proved by experience, that all such societies, when properly conducted, are powerfully instrumental in promoting the advance of science.

Unfortunately, it sometimes happens, that when a new society springs up, it in some degree interferes with a society previously existing. This Association, however, interferes with no other society, and therefore, setting aside the great objects actually accomplished, far beyond the pecuniary resources of other societies, and for which I take no credit, because I presume for a moment they are unknown, it appears to me, nevertheless, to follow irresistibly, that this Association, acting precisely as other learned societies do—using the same means, and exerting a similar indirect influence, must likewise, just as they are, and on a scale just proportioned to its magnitude, be eminently useful in urging on the advance of science.

It may, perhaps, be worth while to inquire for a moment in what way the associations of scientific men promote science. The inquiry, however, cannot alter

[1] Mr. Babbage.

the fact that they do so, for that fact is based on experience. There are many and very obvious ways in which they do so. I shall mention but one.

The love of truth ; the pleasure which the mind feels in overcoming difficulties ; the satisfaction in contributing to the general store of knowledge ; the engrossing nature of a pursuit so exalted as that of diving into the wonders of the creation ; all these are very powerful incentives to exertion ; and under their influence great works have been undertaken in the cause of science, and carried through to a successful termination ; but I believe few will be disposed to deny that further inducements must be highly useful.

Let it be for a moment recollected, that where any, even the most trifling, step in advance has been gained, except perhaps the accidental discovery of a simple fact, there has usually been a long and laborious course of previous preparation. It has been necessary, even in the more popular sciences, to know accurately, first, what had been done by others ; to see distinctly the boundary line between the known and the unknown, before there was the least chance of effecting anything ; and in the higher departments of science such is the time to be expended, so great the toil to be endured in ascending to that elevation, from which the difficulties to be encountered but just begin to appear, that the task is one to which the undivided energies of man exerted for many years are no more than commensurate.

But the necessary preparations accomplished, then the real difficulties commence. Some perhaps apparently new principle suggests itself ; it is followed, with great expenditure of time and labour, to its remote consequences, and it turns out to be perfectly barren and worthless.

One disappointment succeeds another, and years of toil pass away and no result. Under these trying circumstances the associations of scientific men afford their friendly aid ; they soothe disappointment, excite hope, and prepare the way for redoubled exertion ; they call into active existence that principle which has been implanted in our nature for the noblest purposes—the legitimate ambition of meriting and receiving the approbation of our friends and associates. In the ordinary circle of acquaintances, the man engaged in scientific pursuits will find very few, if any, who can understand and appreciate his labours ; but in such associations as this, there are always many who see exactly the object aimed at, the difficulties to be encountered, and who are readily to acknowledge with gratitude every successful effort in the cause of science.

It is thus, without having recourse to other considerations that I account for the fact, that the associations of scientific men, even when they employ no large funds, and perform no gigantic labours, as this Society does, still, by their indirect action, accelerate very greatly the progress of scientific discovery.

But this Association performs other important services. It appears to me to diffuse over scientific inquiry (if I may so express myself) a salutary influence—a healthy vigour of action. What more calculated to dispel that feeling of languor and weariness, the consequence of excessive mental labour long continued, than the freshening excitement of an interchange of ideas with men to whom the same course of research had long been an object of interest ? What more likely to extinguish any petty jealousy which might arise—and scientific men, like other men, have their weaknesses sometimes—than to bring all the parties together in friendly intercourse, where they cannot but feel they have a common object, and are working in a common cause—the discovery of truth ?

Again : should the mind, pursuing in retirement some single scientific object, raise up to itself notions exaggerated and unreal, of the importance of that object, and then, elated and misled by some trifling success, should it throw off the garb of humanity, the characteristic of science pursued in a proper spirit, what more calculated to dispel the illusion than these meetings, where the man, however eminent in that branch of science to which he may have devoted his almost exclusive attention, will be sure to find others immensely his superior in every other department of human knowledge ? And it is not merely for the sake of individuals engaged in the pursuit of science that these consequences are so valuable ; it is also for the sake of science itself.

It is important that science should stand before the world in an aspect which is not forbidding, and we may rest assured of this, that wherever there may be the least trace of petty jealousy, of prejudice, or of pride, the world will not be slow to discover it ; and as science claims as one of its noblest attributes, the power of exalting and enlarging the mind, and of arming it against such weaknesses, it will thus be exposed to the charge of having preferred pretensions to which it has no just title.

I will not detain you by enlarging upon the other obvious beneficial consequences of these meetings, such as the opportunities they afford for the free discussion of questions upon which the concentrated knowledge of individuals may be brought to bear with so much success—the opportunities they afford for the formation of new friendships between scientific men, often fraught with consequences very important to science, and the necessary tendency of them to encourage a taste for science. Upon all these I will abstain from offering any observations. There is, however, one consequence of these meetings, to which, if you will permit me to detain you a moment longer, I will just advert.

It has been remarked by a modern traveller of considerable depth of observation, that he had always found in the children of the fields a more determined tendency to religion and piety than amongst the dwellers in towns and cities, and that he conceived the reason to be obvious—that the inhabitants of the country were less accustomed to the works of man's hands than to those of God. May not the observation be of more extensive application than at first sight appeared ? and if it be true that where we dwell constantly in large cities the mind is liable to be led astray by the habitual contemplation of the works of man, forced upon it imperceptibly by the continual succession of ideas—all of the same character— all originating in objects which have been shaped and fashioned by man, may it not also be true that it is equally liable to be led astray where it concentrates its whole attention, and exerts its whole energy without relaxation in the contemplation of the greatest of all human works, that which the labour of so many centuries has raised up—the structure of the abstract sciences ? And if that be so, what more calculated to unbend the mind, and to divert for a season the current of ideas into other channels, than these periodical meetings, where, in the proceedings of every section, matter will be found of the deepest interest to every true philosopher ; and where, however dissimilar the facts, however varied the inferences, the result will everywhere be still the same—that of putting forward more prominently in bold relief the wonderful works of creation ? It appears to me, if I may presume to offer an opinion on such a subject, that the continual progress of discovery is destined to answer objects far more important than the mere improvement of the temporal condition of man. Were there a limit to scientific discovery, and had we reached that limit, we should be in the condition of a man who, with the most splendid landscape before him, was insensible of its beauty because the charm of novelty had passed away. Each successive discovery, as it brings us nearer to first principles, opens out to our view a new and more splendid prospect, and the mind, led away by its charms, is carried beyond and far above the petty and ephemeral contests of life ; but the more rapid the discoveries are the more powerful the charm, and therefore great is the motive for exertion ; and in labouring in this cause there is this gratifying reflection, that our labours cannot injure our successors, for the religion of discovery is rich beyond the powers of conception ; and however much we may draw from it we shall not leave its treasures exhausted—no, not even diminished, because they are infinite. This Association has already accomplished much ; I feel persuaded it will accomplish much more ; but of this we may rest assured, that however long it may endure, and I see no principle of endurance which other societies have that is here wanting, it will find an ample and an enlarging field of useful employment.

On the Construction of Large Reflecting Telescopes.

By The Earl of Rosse[1]

THE Council having intimated their opinion that some account of the experiments in which I have been engaged on the reflecting telescope would not be altogether devoid of interest, I will endeavour to describe as briefly as possible the manner in which I have attempted to accomplish the object in view, and the principal results obtained.

Having concluded that upon the whole there was a better prospect of obtaining by reflexion rather than by refraction the power which would be required for making any effectual progress in the re-examination of the nebulæ, the first experiments were undertaken in the hope of obviating the difficulties which had previously prevented the application of the brilliant alloy, which may be formed of tin and copper in proper proportions, to the construction of large instruments. The manner in which the difficulty had been met was by adding an excessive proportion of copper to the alloy, but the mirror was no longer susceptible of a durable polish, and when used its powers declined rapidly.

[1] B. A. Report, 1844, 79.

It appeared to me, therefore, to be an object so important to obtain a reflecting surface which would reflect the greatest quantity of light, and retain that property little diminished for a length of time, that numerous experiments were undertaken and perseveringly carried on. After a number of failures, the difficulties appeared to be so great that I constructed three specula, where the basis of the mirror was an alloy of zinc and copper in the proportion of 1 zinc to 2·74 copper, which expands with changes of temperature in the same proportion as speculum metal. This was subsequently plated with speculum metal, in pieces of such size as we were enabled to cast sound. These specula were very light and stiff, and their performance upon the whole satisfactory ; but they were affected by diffraction at the joinings of the plates, and although very brilliant and durable, defining all objects well under high powers, except very large stars, still as the effect of diffraction was then perceptible, they could not be considered as perfect instruments. In the course of the experiments carried on while these three specula were in progress, it was ascertained that the difficulty of casting large discs of brilliant speculum metal arose from the unequal contraction of the material, which in the first instance produced imperfections in the castings and often subsequently their total destruction ; and it appeared evident that if the fluid mass could be cooled throughout with perfect regularity, so that at every instant every portion should be of the same temperature, there would be no unequal contraction in the progress towards solidfication, nor subsequently in the transition from a red heat to the temperature of the atmosphere. Although it was obvious that the process could not be managed so that the exact condition required should be fulfilled, still by abstracting heat uniformly from one surface (the lower one), the temperature of the mass would be kept uniform in one direction, that is, horizontally ; while in the vertical direction it would vary in some degree as the distance from the cooling surface. These conditions being satisfied, we should likewise have a mass which would be free from flaws, and when cool would be free from sensible strain : nothing could be easier than to accomplish this approximately in practice ; it would be only necessary to make one surface of the mould (the lower one) of iron, of a good conducting material, while the remainder was of dry sand. On trial this plan was perfectly successful ; there was however a new, though not a very serious defect, which was immediately apparent ; the speculum metal was cooled so rapidly, that air-bubbles remained entangled between it and the iron surface, but the remedy immediately suggested itself ; by making the iron surface porous, so as to suffer the air to escape, in fact by forming it of plates of iron placed vertically side by side, the defect was altogether removed. It only then remained to secure the speculum from cooling unequally, and for that purpose it was sufficient to place it in an oven raised to a very low red heat, and there to leave it till cold, from one to three or four weeks, or perhaps longer, according to its size.

The alloy which I consider the best differs but little from that employed by Mr. Edwards ; I omit the brass and arsenic, employing merely tin and copper in the atomic proportions, namely, one atom of tin to four atoms of copper, or by weight, 58·9 to 126·4. As it was obviously impossible to cast large specula in earthen crucibles, the reverberatory furnace was tried, but the tin oxidized so rapidly that the proportions in the alloy were uncertain, and after some abortive trials with cast iron crucibles, it was found that when the crucible is cast with the mouth up, it is free from the minute pores through which the speculum metal would otherwise exude ; and therefore such crucibles fully answered the purpose.

It was very obvious that the published processes for grinding and polishing specula, being in a great measure dependent on manual dexterity, were uncertain and not well suited to large specula ; accordingly, at an early period of these experiments, in 1827, a machine was contrived for the purpose which has subsequently been improved, and by means of it a close approximation to the parabolic figure can be obtained with certainty : as it has been described in the Philosophical Transactions for 1840,[1] it is unnecessary to do more than to point out the principle on which it acts ; the speculum is made to revolve very slowly, while the polishing tool is drawn backwards and forwards by one excentric or crank, and from side to side slowly by another. The polishing tool is connected with the excentrics, by a ring which fits it loosely, so as to permit it to revolve, deriving its rotary motion from the speculum, but revolving much more slowly. It is counterpoised so that it may be made sufficiently stiff, and yet press lightly on the speculum, the pressure being about one pound for every circular superficial foot. The motions of this machine are relatively so adjusted, that the focal length of the speculum during the polishing process, or towards the latter end of it, shall be gradually becoming slightly longer ; and the figure will depend in a great measure upon the rapidity with which this increase in the focal length takes place. It will be evident that a surface spherical originally will cease to be so if, while subjected to the action of the polisher, it is in a continual state of transition from a shorter to a longer focus ; in fact during no instant of time will it be actually spherical, but some curve differing a little from the sphere, and which may be made to approach the parabola, provided it be possible in practice to give effect to certain conditions.

An immense number of experiments, where the results were carefully registered, eventually established an empirical formula, which affords at present very good practical results, and may hereafter perhaps be considerably improved. In fact, when the stroke of the first excentric is one third the diameter of the speculum, and that of the second excentric is such as to produce a lateral motion of the bar which moves the polisher, measured on the edge of the tank, equal to 0·27, the diameter of the speculum (or referred to the centre of the polisher to 0·17), the figure will be nearly parabolic. The velocity and direction of the motions which produce the necessary friction being adjusted in due proportion by the arrangements of the machine, and the temperature of the speculum being kept uniform by the water in which it is immersed, there remain still other conditions which are essential to the production of the required result. The process of polishing differs very essentially from that of grinding ; in the latter, the powder employed runs loose between two hard surfaces, and may produce scratches possibly equal in depth to the size of the particles ; in the polishing process the case is very

[1] Pp. 80 *seq.* of the present collection.

E

different ; there the particles of the powder lodge in the comparatively soft material of which the surface of the polishing tool is formed, and as the portions projecting may bear a very small proportion to the size of the particles themselves, the scratches necessarily will be diminished in the same proportion. The particles are forced thus to imbed themselves, in consequence of the extreme accuracy of contact between the surface of the polisher and the speculum. But as soon as this accurate contact ceases, the polishing process becomes but fine grinding. It is absolutely necessary therefore to secure this accuracy of contact during the whole process ; if the surface of a polisher of considerable dimensions is covered with a thin coat of pitch of sufficient hardness to polish a true surface, however accurately it may fit the speculum, it will very soon cease to do so, and the operation will fail. The reason is this, that particles of the polishing powder and abraded matter will collect in one place more than another, and as the pitch is not elastic, close contact throughout the surfaces will cease. By employing a coat of pitch, thicker in proportion as the diameter of the speculum is greater, there will be room for lateral expansion, and the prominence can therefore subside and accurate contact still continue ; however, accuracy of figure is thus to a considerable extent sacrificed. By thoroughly grooving a surface of pitch, provision may be made for lateral expansion contiguous to the spot where the undue collection of polishing powder may have taken place. But in practice such grooves are inconvenient, being constantly liable to fill up ; this evil is entirely obviated by grooving the polisher itself, and the smaller the portions of continuous surface, the thinner may be the stratum of pitch.

There is another condition which is also important, that the pitchy surface should be so hard as not to yield and abrade the softer portions of the metal faster than the harder ; when the pitchy surface is unduly soft, this defect is carried so far that even the structure of the metal is made apparent. While therefore it is essential that the surface in contact with the speculum should be as hard as possible consistent with its retaining the polishing powder, it is necessary that there should be a yielding where necessary, or contact would not be preserved ; both conditions can be satisfied by forming the surface of two layers of resinous matter of different degrees of hardness ; the first may be of common pitch adjusted to the proper consistence, by the addition of spirits of turpentine or rosin, and the other I prefer making of rosin, spirits of turpentine and wheat flour, as hard as possible consistent with its holding the polishing powder. The thickness of each layer need not be more than $\frac{1}{40}$th of an inch, provided no portion of continuous surface exceeds half an inch in diameter ; the hard resinous compound, after it has been thoroughly fused, can be reduced to powder, and thus easily applied to the polisher, and incorporated with the subjacent layer by instantaneous exposure to flame. A speculum of three-feet diameter thus polished has resolved several of the nebulæ, and in a considerable proportion of the others has shown new stars, or some other new feature ; and by the same processes a speculum of six feet diameter has just been completed.

Plain Specula of Silver[1]

THE Earl of Rosse said that, having observed in the London papers that the President of the Association had in his inaugural address conferred upon him the honour of alluding with approbation to the attempts which he had lately made, and with considerable success, to procure plain specula of silver for reflecting telescopes, he thought that perhaps the Section might wish to hear some particulars, and if they could spare a few minutes he would make a short statement on the subject.

It is well known that about one-third of the light which falls upon the great speculum of the Newtonian telescope is lost in the first reflexion, and that nearly one-third of the remainder is lost in the second reflexion. Light being of the greatest importance, especially in the examination of faint objects, in the Herschelian telescope the second reflexion has been dispensed with altogether; and in the Newtonian telescope attempts have been made, by substituting a prism for the flat mirror, in some degree to reduce the amount lost. In the Herschelian telescope the mirror is inclined, so that the light proceeding in a direction parallel to the axis of the tube is reflected to the centre of the eye-glass fixed to the side of the tube; there is thus but one reflexion. A consequence, however, of placing the great mirror obliquely is, that though it may be truly parabolic, yet a pencil of light proceding from a point in an object will not be reflected to a point as it should be, but will be diffused over a certain space. In a telescope of 3-feet aperture, and 27-feet focus, the diameter of that space will be more than $\frac{1}{100}$th of an inch, so that the magnifying power employed cannot be considerable without producing indistinctness. Were it possible to work the surface assigned by theory for oblique reflexion as accurately as the surface of the paraboloid, we should have the light without the indistinctness; but that has not yet been accomplished. Where specula are very large, it has been proposed to place the eye-glass in the axis of the tube, and it has been contended that the light interrupted by the head and shoulders of the observer would be less than the light lost by the second reflexion. This is no doubt true, and various ways have been suggested of placing the observer so that the light interrupted should be a minimum, the temperature of the air in the tube remaining at the same time undisturbed. In such a construction, however, there would be great diffraction, and that appears to me to be an insurmountable objection. The rectangular prism was proposed by Newton as a substitute for the plain speculum; and with a prism of $2\frac{1}{8}$ inches aperture by Mertz, the saving of light is considerable. Mertz informed me that a somewhat larger prism might be made, but that there would be considerable delay: he did not, however, hold out any hopes of being enabled to make one large enough for our 3-feet speculum. It is evident that as the size of the prism is increased, the amount of light lost in

· B. A. Report, 1851, 12 (sectional matter).

passing through the glass will be greater, and a point will at length be arrived at, sooner or later, according as the glass is more or less transparent, when the light lost in prismatic reflexion will be as great as in reflexion from a speculum of metal. It occurred to me that a small prism might be substituted for a large one by placing it near the focus, and that practically the inconvenience of a small field might be obviated to a certain extent by employing a plain speculum and eye-piece in the ordinary way for general work, to be turned aside by a suitable contrivance when the prism was to be made use of. An eye-piece, of course of an unusual construction, would be required, and it does not seem practicable to make such an eye-piece with two lenses achromatic : the four-glass eye-piece would be so, but some light would be sacrificed. How far the want of achromatism would interfere with real work I have not proceeded far enough to be able to say. Achromatic prismatic refraction has been proposed by Sir David Brewster as a substitute for the plain speculum. It has not, as far as I am aware of, been tried, but the great size of the prism which would be required appears to me to be a serious objection.

Under these circumstances, it is obvious that where there was a reasonable prospect of obtaining a material for the plain speculum more reflective than the alloy of tin and copper, no time nor labour could be misapplied in endeavouring to effect so important an object. It was not until Jamin's ' Mémoire sur la Couleur des Métaux ' appeared in the ' Annales de Chimie ' for 1848, that I was aware that silver reflected so very much more light than speculum metal. Jamin states that he has abundantly verified Cauchy's formulæ for the laws of metallic polarization ; and from these laws, by the aid of certain constants, he has determined the intensity of light reflected by some of the metals. From the table he has given, it appears that while speculum metal reflects about sixty rays out of a hundred, silver reflects ninety. Not from having any doubt of the accuracy of Jamin's deductions, but happening to have the means at hand, I made a few coarse photometric measurements of the relative reflective powers of silver and speculum metal, and the results appeared to coincide sufficiently with the more accurate deductions of Jamin. Not feeling very sanguine as to the practicability of procuring an accurate surface of silver by mechanical means, owing to its softness, I tried the electrotype process in the first instance. The silver was thrown down upon a surface of highly polished speculum metal ; but in every case where the speculum metal was thoroughly clean there was strong adhesion, so that separation could not be effected without destroying both surfaces. The means which were employed to guard against adhesion in electrotyping the engraved copper plates for the Ordnance Map of Ireland, which, in fact, consisted in applying an extremely thin film of wax, would obviously be inadmissible in this case. Several attempts to precipitate silver on a steel speculum failed, as the silver had not a proper polish. Being however without experience in electrotyping, I do not consider these failures conclusive. Attempts were made with a steel die truly polished to procure accurate polish by pressure. This did not succeed : the surface was not sufficiently true, owing apparently to unequal elasticity in the texture of the silver. To turn or plane a polished surface was not attempted, as the idea had suggested itself to the late Mr. Barton, and if it had been practicable in his hands it would doubtless have succeeded. There remained therefore but to try some modification of the ordinary processes of polishing metals. A difficulty occurred in the beginning, which gave considerable trouble. Owing to the softness of the silver, the emery employed to grind it flat became imbedded in it, and in that state to polish it was quite impossible. Without grinding, however,. there seemed to be no means of making the silver sufficiently flat, as Whitworth's scraping process had been tried, and was not found to be sufficiently delicate. The difficulty was obviated by employing a bed of German hones, by which the silver is probably rather filed than ground. The blue hones might perhaps be employed with advantage after the German hones, but I have not tried them. The silver being thus prepared, I tried the ordinary process of polishing on pitch in vain, employing the pitch of various degrees of hardness ; the surface of the silver was irregular and the polish imperfect. The silversmith makes use of chamois leather and rouge, sometimes finishing with the hand charged with rouge ; his polish is very brilliant, but the surface is, as we should expect, very untrue, as for example, the surface of a highly finished plateau. It was rather puzzling to find that while chamois leather charged with rouge polished silver, pitch, however soft, did not ; there was no apparent difference in the two cases, but this ; that the pitch slightly shielded by the rouge came more or less in contact with the silver, while the chamois leather, holding fast the stratum of rouge, was scarcely in contact at all with the silver. That this was the true reason was probable, as in proportion as the chamois leather was less shielded, the process was imperfect. It is evident that, however fine the polish, a true surface could not be procured by the action of an elastic material, and therefore pitch was taken as a basis ; a substance which is solid, and at the same time adapts itself in the most perfect manner to the surface to be polished. We proceeded in this way :—Pitch of the proper hardness for polishing speculum metal was covered with a mixture of rouge and the combination of ammonia and soap which we employ in polishing specula. The silver was worked upon this for a short time to force the rouge into the pitch, after which the rouge mixture was again applied and suffered to dry. The surface was then slightly moistened with spirits of turpentine, and more of the rouge mixture applied. The following day, the turpentine having evaporated, there was a rouge surface, perhaps $\frac{1}{100}$th of an inch thickness, upon which the silver was polished, using fresh rouge and ammonia soap just as if it was speculum metal. The spirit of turpentine had long been exposed to the light, and consequently was slightly adhesive. Silver was several times polished on this surface successfully, and a plain silver speculum so polished performed well, giving perfectly sharp images. I have been thus minute in explaining what I believe to be the rationale of the process as a guide to others, because, having as yet practised it but little I may perhaps have omitted to notice some things apparently non-essential, but which are really not so.

Drawings to illustrate Recent Observations on Nebulæ[1]

By the Earl of Rosse

With Remarks by Rev. Dr. Robinson

DR. Robinson stated that he had examined the drawings, which contain careful delineations of several nebulæ not previously examined, and certainly the contemplation of them was well fitted to increase the obligations of the astronomical world to Lord Rosse, as well as to fill every mind with astonishment at the wondrous revelations of his matchless telescope. Each of them was a new proof of a former statement of his, that this instrument would probably disclose forms of stellar arrangement, indicating modes of dynamic action never before contemplated in celestial mechanics. He referred to the drawings of M. 51, in which the spiral or vorticose arrangement of the stars and unresolved nebulæ was first remarked in its simplest form ; and to others already published, where it presents itself under conditions of greater complexity. He also referred to the important fact that the class of planetary nebulæ might now be fairly assumed to have no existence, as all of them which have been examined prove to be either annular or of a spiral character. Thus M. 97, which was considered by Sir J. Herschel the finest specimen of them, and seemed even in his 18-inch reflector a uniform disc, presents in the six-feet a most intricate group of spiral arcs, disposed round two starry centres, looking like the visage of a monkey. Among the new ones are H. 2241. It is a ring of stars with a faint nebula within, and a fine double star near its edge ; H. 2075, of the same kind, but with a bright star almost exactly central, and nine others round it, evidently part of the same group. H. 450 is a most extraordinary object ; the ring exactly circular, its light mottled and flickering, and within it what is evidently a globular cluster. Scarcely less surprising, but more magnificent from its association, is the planetary at the edge of M. 46, which he had seen, though in a night not so favourable as that must have been when the drawing was made. It is a resolvable double ring, rather spiral, with a central star ; and from the improbability of two objects so rare as a splendid cluster, and one of these compound rings being *casually* connected, it seems reasonable to think they constitute one system. The double star, ι Orionis, belongs also to this class, and he called attention to the absolute darkness of the aperture in the nebula round the two stars and that the larger of them was at its edge instead of being central. He argued, from the remarkable difference between these objects as seen in the telescopes of Lord Rosse (even in the three-feet) and those of previous observers, how desirable it was that a complete review of the nebulæ should be made without loss of time. Even now much labour and talent were expended in theorizing on the imperfect

[1] B. A. Report, 1852, 22 (sectional matter).

data given by instruments, which though matchless in their time have now been surpassed. Among others he directed the notice of the section to H. 604, where the two clusters and the associated spirals are projected into ellipses ; and to H. 2205, in which the long-resolved ray, being the most intense, was alone seen by Herschel, but the magnificent spirals and their central stars escaped him. M. 65, H. 857, appear to be helices seen obliquely. But the most curious one is M. 33, of which the centre is a triple star disposed as an equilateral triangle among a mass of smaller, from which proceed eight or nine spirals ; and round all is an enormous nebula, in which, however, no spiral character had yet been traced.

There were several examples of another singular system, nebulæ streaked with dark bands, such as Bond discovered in the great nebula of Andromeda. H. 399, a wisp ; H. 1393, a long ray of most marvellous appearance ; H. 218, an oblique with sixteen or seventeen dark transverse stripes ; and H. 315, having the nebula a cluster nearly insulated by offsets from the broad curved dark band, are among the most surprising. But the number of these curious objects was so great that time would only permit him to invite attention to H. 1052 and 1053, where the cause of spirality had been interrupted by some other forces that bent the system at a right angle and drew the nebula into a straight ray ; to H. 444, a double re-solved nebula inclosed in a large and faint oval ring ; and above all to M. 27, the "Dumb Bell" nebula as shown by the six-feet, with its brilliant two clusters of comparatively large stars, its dark bands and the faint rings which surround it differing even more from the picture of the three-feet than that does from the figure of Herschel.

In the name of the Section he thanked Lord Rosse, not merely for the pleasure which they received from the sight of these wonders, but for the unremitted and precious gifts which he was conferring on astronomy. Would he also increase their gratitude by mentioning any improvements which he might have lately made in the methods of suspending these large specula in their tubes or in the process of polishing, the latter with reference to the possibility of its being prac-tised with success by persons who had not the long experience and mechanical knowledge of his Lordship ?

Lord Rosse adverted to the peculiar conditions of equilibrium which must prevail in these systems, or rather to the forces which are required to produce the peculiar constitution which they indicate, and pointed out the difficulties of such an investigation. It could however not be undertaken with advantage till we possess a much more extended collection of data, to which he would contribute to the utmost of his power. These drawings were based on measures carefully taken with a bar-micrometer (the only one available in such cases), and he believed they might be trusted. He had already described the improvement effected by supporting the speculum on its lever by eighty-one balls, and mentioned the striking fact, that with a speculum weighing $3\frac{1}{2}$ tons a slight pressure of the hand would deform for a time the image of a star. He had since effected a further improvement by supporting the edge of the speculum in a hoop mounted in jimmals. As to polishing, he had recently made many experiments with 3-feet specula in reference to the object of Dr. Robinson's question, and in particular had found, that, by increasing the speed of the second excentric in his machine, the process was rendered so much more certain, that desiring one of his workmen, a smith, to perform the whole process *without any superintendence* on his part, he produced a speculum, not perhaps absolutely perfect, but capable of doing excellent work. He had no doubt that any person of ordinary mechanical capacity would be able to do as much with a little instruction, and he would be most willing to give that instruction to any observer that might be placed in charge of a large reflector.

First Report of the Committee, consisting of the Earl of Rosse, the Rev. Dr. Robinson, and Professor Phillips, appointed by the General Committee at Belfast, to draw up a Report on the Physical Character of the Moon's Surface, as compared with that of the Earth[1]

I.   The Committee, having received their instructions in September, 1852, lost no time in assembling, by invitation of the Earl of Rosse, at Parsonstown, where with the assistance of Colonel Sabine, at that time President of the Association, they made preliminary examinations of the moon, by the powerful telescopes of the Earl of Rosse, and formed plans of further proceeding in conformity with the results of these examinations, and the individual experience of the members of the Committee.

II.   Taking as a general basis for the work to be done, the much-valued maps and treatise of Mädler and Beer, it appeared to the Committee desirable to procure a new set of drawings or surveys of selected parts of the lunar disc ; to suggest certain conditions of representation, with reference to the illumination of these parts, and to propose a uniform scale for the drawings.

The suggestions offered, as some help to observers on this subject, were the following :—

" 1. For the acquisition of correct ideas regarding the form of any part of the lunar disc, an examination of it under at least three aspects appears indispensable.

    a. A little (one hour?) after the sun rises on that part of the spherical surface.

    b. When the sun is on the meridian of that part.

    c. A little (one hour ?) before the sun sets upon it.

" By this arrangement each part of the surface may be delineated and described under three directions of incident sun-light, two of them (*a* and *c*) suited by long shadows to discover the inequalities of level, and the other (*b*) aiding by a vertical incidence to make apparent the unequal reflective powers and different colours which characterize the different lunar regions, and the systems of brilliant stripes which are connected with certain lunar forms.

[1] B. A. Report, 1853, 84.

" 2. The age of the moon,' when a drawing is made, should be stated to the second decimal of the day, because a knowledge of this epoch is essential to a right estimation of the angle of incident light under which the observations are made. Probably the observer will find it convenient to prepare beforehand a table of the moon's age, corresponding to each hour of mean solar time. The mean solar time of the place at the beginning and end of each observation should also be stated.

" 3. Among the chief points to be attended to are—

a. The steepness of slopes, which may perhaps be best determined by noting the time at which they began or ceased to be illuminated generally.

b. In ring mountains the difference of level between their exterior and interior bases.

c. The curvature of their interior, whether greater or less than that of the general surface. Some of them are much raised in the centre, as is evident by the shadows which these parts throw.

d. Whether the brilliant stripes are elevated above the ground where they pass, and the angle of illumination at which they disappear.

e. Stopes, height, and breadth of the soft ridges in the Maria.

f. External fragments round ring mountains.

g. Relation between mass of wall and area of depression (*i.e.*, would the wall fill up the hollow).

" 4. In delineating the appearances on the moon's surface, the Committee think the observer must be encouraged to employ various methods. For a general view of the proportional areas, more accurate than any sketch, Photography may be employed. To steady the work, and reduce it to a uniform scale, micrometrical measures will be required. In some cases, where these cannot be supplied by the observers separately, they may be obtained at one of the observatories. In drawing by the eye the Camera lucida is available, if the telescope has an equatorial movement by clock—a condition not only desirable, but perhaps indispensable for perfect delineation.

" 5. For convenience of comparison, it appears desirable to recommend that one uniform scale should be employed in the delineations. Though it may seldom be practicable to employ on the moon a power of 1000, the Committee recommend that the *drawings* should in no case be made on a smaller scale. If the distance of the paper from the eye be assumed at 10 inches, a circular space on the moon's surface one mile across will be represented with only a diameter of about one-twentieth of an inch. For objects which require larger representation, the ordinary scale may be doubled or tripled.

" 6. Both in drawing and describing it appears desirable to employ the method of Beer and Mädler, who *draw* the moon as she appears in the inverting telescope, but *describe* the relative situation of her parts by reference to her poles as northern and southern, and sides as east and west, in correspondence to the nearest cardinal points of the earth.

" 7. It is found by trial that Sepia drawings are well suited for representations of the peculiarities of the moon's surface."

It is requested that the drawings and descriptions, which may be prepared in conformity with these suggestions, may be forwarded by post to Professor Phillips, Assistant-General Secretary of the British Association, St. Mary's Lodge, York.

III. The Committee next endeavoured, by circular, to obtain the co-operation of a limited number of gentlemen, whether in the British Islands or in foreign parts, who by their possession of instruments of adequate optical power, habits of astronomical observation, and available leisure, might be able and willing to undertake definite parts of the great task which they hoped to see accomplished.

IV To these letters, the replies which have been received offer in general very satisfactory assurances of co-operation ; and in some cases useful additional suggestions and notices of interesting facts are added.   In particular, the author of " Der Mond," besides assuring the Committee of a general desire to co-operate in their labours, states the degree in which, since his appointment to the Observatory at Dorpat, he has been able to extend his former observations on the " light streaks " of the moon, an object to which the Committee had ventured to specially direct his attention, and instances the distinction which he has already made between the " light spots " which vanish in lunar eclipses, and those which remain visible and even grow more distinct in the shadow, except where it is deepest.

The Committee do not, however, feel it to be proper now to report the special views and limited progress of their members, beyond placing before the Association one drawing of the Mountain Gassendi—on the scale proposed for the whole survey—made from a telescope mounted at York by one of their members.

[The subsequent reports of this Committee did not deal with work at Parsonstown.]

Mechanical Science[1]

Address by Lord Rosse, the President of the Section.

LORD Rosse commenced by apologizing for any oversight he might commit, as he had never at any previous Meeting of the British Association presided over the Mechanical Section. He was happy, however, that there could be no danger of serious errors on his part, as there were able men on both sides of him who had made Civil Engineering their special study. He proceeded to say that the question had sometimes been asked why a Mechanical Section was necessary; might not all mechanical questions be conveniently discussed in the Mathematical and Physical Section ?   To that question, on some occasions, an answer has been returned. It may at once be said, that it has been found eminently useful to have separate Sections for each distinct department.   It is only under such an arrangement that discussion can be really effective in bringing out new truths. If a considerable portion of the Section is not intimately acquainted with the subject

---

[1] B. A. Report, 1857, 175 (sectional matter). This was then the title of one of the Sections of the British Association.

in all its details, what prospect can there be of new and sound views being elicited; and indeed if the whole Section has not some general knowledge of the branch of science which has been committed to its care, what hope can there be that discussions will be heard with interest, and will have real efficiency in awakening and strengthening a taste for science ?   This has been felt, indeed strongly felt, at the Royal Society, where there are no Sections, and the subjects are of a very varied nature, comprehending the whole range of the Mathematical and Physical Sciences, and all the natural Sciences.   A paper perhaps is read on Pure Mathematics : very possibly there may not be in the room at the time more than two or three persons who are intimately acquainted with that branch of science : a discussion of course is out of the question.   A paper follows on one of the Natural Sciences : if there are a few who are working in that direction a discussion takes place ; but it is of little interest even to those engaged in other branches of Natural Science ;   and almost, if not altogether without interest to the Mathematician, the Chemist, the Astronomer, the Geologist, and the Physicist.   So it rarely happens that there is a discussion at the Meetings of the Royal Society, of general interest or of real value ; and for that there is no remedy.   Here, by the happy expedient of breaking up the Association into separate Sections, the way has been prepared for discussing subjects in that effective manner which, originating with the Geological Society, has already so much advanced geological science.

Where one of the great objects of these meetings is to elicit truth by discussion, it is evident how unwise it would be to group together in one Section a variety of subjects, each requiring special studies, a special line of thinking, and special experiences.   In Section A, human ingenuity and human knowledge are employed in the solution of mathematical and physical problems ; while in Section G, human ingenuity and human knowledge, but of a different kind, are employed in the solution of questions of practical engineering.   This may so far perhaps be considered, in one sense at least, a sufficient answer to the question, why is a Mechanical Section necessary ?   The question, however, may be put in another sense. Where the investigations are not abstract, but practical, and where the results generally are of immediate interest, is it necessary that the British Association should interfere at all ? Will not private individuals, from motives of self-interest, devote themselves to the pursuit of Civil Engineering in its higher branches, without any adventitious stimulus ? Will not public men, seeing that the interests of the State, both in peace and war, are bound up with the full development of the resources of engineering, make it their business to acquire such a general knowledge of the subject as will enable them to ascertain when and where to apply for aid in time of difficulty ? The reverse unfortunately is the case ; experience has shown that men, whether in their private or public capacity, do not act in these matters exactly as we should expect ; they do require both to be aided and urged forward.

In this eminently practical country, private individuals, very often relying on experience, neglect the means necessary to render calculation effective. Experience however, is not always at hand, and is often very costly.   How often do we see the ingenious mechanic working on false principles, vainly perhaps attempting to accomplish something which a little elementary knowledge would have shown to be impossible! There are perhaps few gentlemen present who could not point out instances where individuals had sustained heavy losses from the want of adequate theoretical knowledge. In his limited experience he had known several. This perhaps is a striking one. Some years ago he was invited by a physician of eminence in London to visit the works of an ingenious mechanic, who was endeavouring to employ air heated by gas as a prime mover. The physician had embarked £12,000 in the project ; a lady of wealth had speculated in it to the extent of £30,000 ; and various individuals had advanced sums altogether to a large amount. At the entrance of the premises there was the wreck of a gigantic machine of unknown construction : other machines in a dilapidated state were lying about in all directions. It appeared from the explanation of the mechanic, that these huge masses of ruined machinery had been constructed partly for the purpose of ascertaining facts to be found in every elementary treatise, and partly for the accomplishment of objects manifestly impossible. In the construction of the engine itself there was a striking display of great ingenuity constantly engaged in a struggle with the laws of nature. It was perfectly evident that the whole was fated to end in disappointment ; still the mechanic and his patrons, undismayed by repeated failures, and heedless of warnings, which, where there was no science, were without force, struggled on till the project came to an end from exhaustion. Some of the parties were ruined, while all lost the capital they had embarked in the speculation.

One of the objects of the Mechanical Section is to prevent such disasters, and no doubt to a certain extent this has been effected. Another object has been effected also : the importance of engineering science in the service of the State has been brought more prominently forward. There seems, however, something still wanting. Science may yet do more for the navy and army, if more called upon.

A few years ago, in sailing through the harbour of Portsmouth, as the boat proceeded along, the sailors gave a little history of each ship laid up there ; they said : That ship has been but once to sea, and it rolled so it was almost impossible to keep masts in her ; she is not likely to go to sea again. There is another ship which sails so badly that she can neither chase nor run away. There is a ship which can scarcely beat to windward, and if it was blowing hard upon a lee-shore, she would have but little chance. Other ships had other defects. Strange uncertainty ; who could avoid asking the question : Is naval architecture really guided by science ? About that time a little book came out which solved the mystery. It is called 'Lectures on the results of the Great Exhibition : the lectures are by first-rate men. In it there is a lecture by Captain Washington, " On the progress of Naval Architecture," an officer of high scientific attainments, now Hydrographer to the Admiralty. After mentioning the well-known historic fact, that during the late great war our best ships were copies, and not always very successful ones, of foreign models, he proceeds to say, that all who served in the blockading fleets were painfully alive to the fact that our ships were inferior to those of France and

Spain in speed, stability, and readiness in manœuvring. That much loss of life might have been spared if our ships had been in form more on a par with those of our opponents. He attributes their inferiority to the fact, that while in France and Spain, and other continental countries, the aid of science had been called in, and the greatest northern nations had turned their attention to ship-building, the only English treatise at all of a scientific character was published by Mungo Murray, who died a working shipwright. That England has not to this day one original scientific treatise on Naval Architecture. He further states, that of the forty-two men who were educated in the School of Naval Architecture which had been established in 1811, and after a few years suppressed, but five had to this day risen to stations of responsibility, and that the sight might have been seen of men familiar with the differential calculus, chipping timber in the dockyards in company with common mechanics. Cruze, in his article on Naval Architecture in the 'Encyclopædia Britannica,' makes similar statements.

It has been objected, however, that the powers of engineering science have been overrated ; that they had been brought to the test during the late war and had but little strengthened our hands. People seemed to think that scientific invention should have carried all before it. Inventions, however, do not come forth at our bidding ; and are we sure there has been much to attract highly cultivated inventive powers to the science of war ? Have we never heard a whisper of official prejudices and official discouragement ? Moreover, if you invent, the invention soon falls into the hands of a vigilant enemy and you have achieved nothing.

It was not by little inventions that the engineering powers of England could have been brought to bear effectually in the late war. If, when war was imminent, civil engineers had been consulted in conjunction with military engineers and naval men, means perhaps would have been found by which the gigantic engineering resources of this country would have been rendered available. It was going a little too far when it was said that Cronstadt could have been taken by contract. Of this, however, there could have been no doubt, that a certain thickness of wrought iron would have resisted the heaviest ordnance then in use ; that the sea could have carried the weight ; and that no stone walls could have long resisted the close fire of large guns. Moreover, there were actually French experiments made a few years before, which, in the absence of new experiments, would have afforded tolerably accurate data for the necessary calculations.

Let it not be said that engineering science was almost powerless in the late war, till it can be shown that it had a fair trial—that its aid had been called for at a proper time and in a proper manner.

It is scarcely necessary further to insist upon the importance of the Mechanical Section. It is obviously the interest of public men, no less than of private individuals, to pay increased attention to Mechanical Science. If this Section can contribute ever so little to bring out new facts, or to direct attention to facts already known, it will have rendered good service.

[Lord Rosse's interest in naval construction is further illustrated by the correspondence included in this collection in pp. 207 *seq.*]

Mathematics and Physics[1]

Introductory Remarks by the President, the Earl of Rosse.

IT has, I believe, been usual, at least recently, in opening the proceedings, to give as far as may be practicable, a general outline of the business to be brought before the Section, and some kind of notice of the order in which it is likely to be taken. As, however, many papers are often sent in after the meeting of the Section, and as frequently circumstances arise rendering it necessary to alter the order of proceeding, any notice that can be given must be very imperfect ; the daily notices, however, will in some degree supply the deficiency. It has also been usual, I believe, and it is obviously convenient, in some degree to define the general character of the business to be transacted, so that new Members may be enabled better to decide whether to attend this Section or some other. I have made inquiry, and find that already there have been received papers on pure mathematics, applied mathematics, magnetism, light, electricity, and meteorology, besides papers on the construction of philosophical instruments. From the titles of the papers, some idea may be formed of the general character of the business to be transacted ; still there are many subjects, in fact several branches of science, which are as yet unrepresented in the papers.

First as to the papers on pure mathematics, I need perhaps hardly say that essays on so abtruse a subject cannot be of very much interest except to mathematicians ; and even mathematicians, unless the papers happen to relate to the particular branches of mathematics with which they are most conversant, may perhaps be sometimes unable to do more than catch the general scope and leading principles of the paper ; still without mathematical knowledge many may often, in the results announced, and indeed in the remarks casually elicited, obtain interesting glimpses into the nature of mathematical processes, and some idea of the progress making in that direction.

In applied mathematics there is much more of general interest, and the results are often perfectly intelligible without a special education. I recollect at the Meeting of the British Association at Oxford, the general results of a very abstruse investigation in applied mathematics in physical astronomy, were so brought forward as to rivet the attention of the whole Section. It was an account given in general terms by M. Le Verrier of his researches for the identification of a comet.

The discoveries in electricity, magnetism, heat, and light cannot fail to be of great general interest. To the human mind nothing is so fascinating as progress. It is not that which we have long had we most value, but that which we have recently acquired : we especially prize new acquisitions, while we enjoy almost unconsciously gifts of far greater value we have long been in possession of. This

---

[1] B. A. Report, 1859, 1 (sectional matter). The title is that of one of the Sections of the British Association.

is our nature ; thus we are constituted ; it certainly is not surprising therefore that we should have a peculiar relish for new discoveries. The interest of a discovery is not usually confined to the discoverer, unless he is very churlish, or even to those who are endeavouring to discover ; but it often extends to the whole civilized world. The interest is, however, not lasting ; for a time we are dazzled by the brilliancy of the discovery ; gradually, however, the impression becomes fainter, and at last it is lost entirely in the splendour of some fresh discovery, which carries with it the charm of novelty. When we reflect upon this, we cannot but perceive how very different the state of the world would have been had mankind from the beginning been in possesion of all the knowledge we now have, and there had been no progress ever since. We ask, why have all these wonders been placed before us—hidden, veiled—only to be brought to light by the vigorous use of our faculties ? How wonderful from its origin has been the progress of geometrical science ! Beginning perhaps 3000 years ago almost from nothing, one simple relation of magnitude suggesting another ; the relations becoming gradually more complicated, more interesting, more important, till in our day it expands into a science which enables us to weigh the planets ; more wonderful still, to calculate long beforehand the course they will take acted upon by forces continually varying in direction and magnitude. When we ask ourselves such questions as these considerations suggest, and thoughtfully work out the answers as far as possible in their full depth of detail, we become in some degree conscious of the immense moral benefits which the human race has derived, and is deriving, from the gradual progress of knowledge. The discoveries, however, in physical science are often immediately applicable to practice, giving man new powers, enabling him better to supply his many wants. We therefore, who are all, in some degree at least, utilitarians, on that account very naturally regard them with deep interest. I am sure the mere mention of the subject has already suggested to you many of the extraordinary discoveries of latter times ; for instance, the production of force almost without limit by heat, and its application to locomotion by sea and land— the transmission of thought, not slowly by letter, not to short distances by sound, but instantaneously to immense distances by electricity ; and when we look around us and see how man has appropriated to his use the properties of light and heat, the powers of wind and water, the materials which have been placed before him in endless variety on the surface of the globe which he inhabits,—that he has effected all this by knowledge accumulated by what we call Science,—it is surely not surprising that we should look upon new discoveries with surpassing interest. The mere utilitarian, however, has been often reminded that discoveries the most important, the most fruitful in practical results have frequently in the beginning been apparently the most barren, and therefore that the discoveries in abstract science are not without interest even for him.   I confess, however, that the gradual development of scientific discovery—in fact, in other words, the steady flow of knowledge into the world—which like a stream becoming broader and deeper as it proceeds points to its own source, to its own origin, which is the origin of man— I confess that this arrangement appears to me to serve far nobler purposes than merely to minister to the corporeal wants of man, as they increase, or are proposed to increase, with the progress of civilization. What those purposes are, to some extent, I think we may clearly see, though to fathom the full depth of such an inquiry would be beyond our powers. Looking merely on the surface, we perceive that the continual springing up of new facts, new discoveries, in endless succession the rewards of industry, must tend to make man industrious. It inspires him with hope, entices him to labour with his mind—the hardest of all labour ; it quickens his faculties, it forces him to look behind and before, to the past and future, and it promotes in him a high moral training by the influence it exercises over his habits and thoughts. Many, no doubt, will feel anxious to see principles immediately applied to practice ; in common language, to see principles made useful ; they will be highly gratified in the Mechanical Section. Here they may, perhaps, occasionally see the same thing ; but more frequently they will find that the results are but stepping-stones which prepare the way for further progress. These few remarks, which I have made principally for the convenience of new Members, will, I think, be sufficient to give some little idea of the kind of business to be transacted here, and I will not allude to the actual practical results which have immediately followed from the labours of this Section. They have been detailed, and recently, especially by my friend on my right hand, Dr. Robinson ; and I will only further add, that I feel much gratified to find so large an attendance of eminent men of science here, ready to correct oversights and supply deficiencies. They, I am well aware, are far more competent to preside here than I can be ; but, with their assistance, the duty will be light ; and as the Council, no doubt on good grounds, have made the present arrangement, I will, without hesitation or mis-giving, at once proceed with the business.

From
MONTHLY NOTICES OF THE
ROYAL ASTRONOMICAL SOCIETY

Notes on Experiments relative to Lunar Photography and the Construction of Reflecting Specula.

By The Earl of Rosse[1]

(Extract of a Letter from the Earl of Rosse to the Astronomer Royal).

"As you mentioned to me in one of your letters that the Astronomical Society would be glad to hear from time to time what we are doing, there are two or three little matters I have been recently engaged in which may, perhaps, interest them.

"First, as to lunar photography. I have constructed a smooth motion-clock to carry the plate of glass, and its performance is satisfactory. The regulator is thus made : there are two levers with balls on the extremities, which exactly balance in 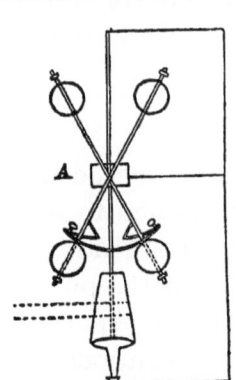 *every position ;* they are acted upon by two springs with screw adjustment, and on the expansion of the balls the regulating friction takes place at the ring A.[2] The object was to obtain a regulator independent of position. The direction of the motion of the glass plate is regulated by an adjustable slide, and we set the slide by trial, not by a table computed for the purpose. To set the slide, an eye-piece with lines truly parallel to the slide, is inserted. By such means a pretty picture of the moon can be obtained, but at present I believe there is no known photographic process which is sufficiently sensitive to give details *in the least degree* approaching to the way in which they are brought out by the eye. The application of such a smooth motion-clock to instruments *not equatoreally* mounted, may, perhaps, be important, as it affords great facilities for the use of the micrometer. With our 3-foot telescope, I have no doubt, excellent micrometric measures might be obtained ; and with a somewhat enlarged small speculum, there would be ample time without hurry.

---

[1] Monthly Notices R. A. S., XIV, 199 (1854).

[2] The crossed rods pass in the inside of a ring and rub the inner edges of the ring.

F

For all objects but the moon, a table might be constructed with little trouble for setting the slide, which would save time.

"You recollect, no doubt, how greatly superior silver would be to speculum metal, it if could be as well and as easily polished as speculum metal. At the Ipswich meeting of the British Association I described a process which had been, to a certain extent, successful. It is difficult, however, and uncertain; and as a silver surface is very perishable, it would scarcely be worth while to employ it, except under special circumstances.

"Another method which I have very recently tried is perfectly easy, and promises well. A plate of glass is coated with silver by precipitation from saccharate of silver. The silver film is then varnished with tincture of shell-lac, and when dry the temperature of the glass is gradually raised to the fusing point of shell-lac. Pieces of shell-lac are then laid upon it, and over them a piece of thick glass. A slight weight presses out the superfluous shell-lac, and the whole having gradually cooled, the silver film adheres permanently to the shell-lac, the glass upon which it had been originally precipitated being easily removed without injuring it. We have thus a silver surface apparently as true as the glass upon which it had been precipitated, and with a beautiful polish. The experiment is imperfect as far as this, that as yet merely common plate-glass has been tried, and not a true glass surface; and as I am about to set out for London, I shall have no opportunity for some time of completing these experiments.

"With the view of applying Mr. Lassell's levers to one of our 6-foot specula, should there be a reasonable prospect of improving its performance in that way, I have tried some experiments as to the practicability of drilling speculum metal. I find it can be drilled by a tubular drill of soft iron and emery, the core being from time to time removed by a pointed chisel and a very light hammer, by which it can be safely broken up gradually. A drill with diamonds set in a groove cuts it well also; and even a drill of perfectly hard steel, revolving slowly, cuts it well; so that there can be no serious difficulty in making the necessary perforations.

"*Castle, Parsonstown, April 20th*, 1854."

The Astronomer Royal having resigned the chair to Mr. Sheepshanks, proceeded to give a full explanation to the meeting of the various experiments alluded to in the foregoing communication from Lord Rosse, his remarks being rendered readily intelligible by means of models, which he had caused to be constructed for the purpose of illustration, and which had been obligingly forwarded by him to the Apartments of the Society. He pointed out the advantage which the system of supports for resisting edgewise pressure, now proposed by Lord Rosse, would have over that of Mr. Lassell, of which it was a modification, namely, that in the case of a reflector being mounted equatoreally, it would prevent the possibility of undue pressure against the *side* of the supporting ring during the period when the telescope was being pointed to a celestial object; whereas in Mr. Lassell's system the supports come into their proper operation only after the position of the speculum has been rectified by rotating the tube in its cradle: this is a point of great importance in large specula, as it is found that they do not immediately recover their normal figure after distortion by pressure. At the same time, he took occasion to repeat that he still thought it preferable to avoid the rotation of the telescope tube in its cradle altogether, which the altazimuth mounting he had proposed would effect, and yet still provide for an equatoreal movement in the telescope, although he was aware that Mr. De la Rue was of opinion that the difficulties of mounting very large telescopes on an equatoreal stand might be overcome. The Astronomer Royal also pointed out, that if such a mounting were adopted, it would be desirable to so construct the system of supports for resisting the pressure perpendicular to the surface of the mirror, as to admit of the fulcra of the levers for resisting edgewise pressure being carried by them, in order that the ends of the levers might support the pressures in all directions equally, without constraining the mirror in any way, and without impeding its change of position by any sensible friction.

Description of an Equatoreal Clock.

By Lord Oxmantown[1]

THE following description of an apparatus for giving an equable motion to a heavy Equatoreal of 18 inches aperture, may, perhaps, be interesting to some of the Fellows.

It was constructed last autumn, and its merits, if it has any, is that it can be very easily made. No nice workmanship is required ; a joiner and plumber can execute the work with sufficient accuracy.

The motive power is a piece of wood closely fitting a wooden box containing water, on the surface of which this piece of wood floats. This float is covered with canvas saturated with pitch, to prevent it from imbibing water, and so expanding and sticking fast in the box.

It has also pieces of sheet-brass fastened on the edges over the pitched canvas, to prevent the pitch adhering to the sheet-lead with which the box is lined. A tube from the bottom of the box, terminating in a piece of flexible pipe, the extremity of which is kept at a uniform depth below the surface of the water in the box by being suspended from the float, allows the water to escape at a constant rate which is regulated by a valve in this pipe. $(h\,h)$ is the box ; $(i\,i)$ the float, which is attached to a brass tube $(d\,d)$, which acts as a guide by passing through a hole in the cross-piece of wood $(f\,f)$, supported by two upright pieces $(g\,g)$; $(n\,n)$ is the India-rubber tube, through which the water escapes ; (o) the valve to regulate the flow ; (m) is a cock, which is closed when the clock movement is not required ; $(q\,q)$ a second box to receive the water as it flows out of the first. When we wish to employ the clock, we, first, pump up water from (q) to (h) through (p)

[1] Afterwards Fourth Earl of Rosse. This description, from Monthly Notices R. A. S. XXVI, 265 (1866), is referred to in the paper from Phil. Trans. 1867, hereinafter reproduced.

and (l) ; secondly, we get the object to be observed into the centre of the field ; thirdly, open the cock (m) and clamp the sector (t) to the polar axis (z) by means of the screw (u) ; and if the object is not now sufficiently in the centre of the field, it may be brought there accurately by turning the nut (c), which draws the sector (t) further from or nearer to the float, by bending the spring (w).

The accuracy of this clock depends, to a great extent, upon the care which is taken to make the sides of the box parallel and flat, so that the horizontal section may be the same at all levels, and to fit the float to it carefully, so that the interval between the float and the sides of the box may be as small as is consistent with absence of friction. Let A be equal to the area in square inches of horizontal surface of float, and $\dfrac{A}{n}$ area of interval between the float and sides of the box ; then, if the float be displaced $\dfrac{1}{m}$th of an inch from the position of the equilibrium, surface of the water round the float will be displaced through $\dfrac{n}{m}$ inches ; therefore if $w =$ weight in lbs. of 1 cubic inch of water, the force tending to bring the float back to its former position $= w.\ A\left(\dfrac{n}{m} + \dfrac{1}{m}\right)$; whereas, if the float had rested on a surface of water of infinite extent, the corresponding force would have been w A $\times \dfrac{1}{m}$ the ratio of the forces in the two cases $= n + 1 : 1$. In the case of the instrument to which this apparatus has been applied, A $= 1296$, and if, for example, the interval between the float and sides of the box be $\frac{1}{16}$ of an inch, and if the variation of friction of the polar axis be taken at 3 lbs. (the average friction being about 7 lbs.), the displacement of the float, which would have been equal to ·064 of an inch on a surface of water of infinite extent, will become equal to

$$·064 \times \frac{1}{1 \times \dfrac{36^2}{4 \times 36 \times \frac{1}{16}}} = \frac{·064}{145} = ·00044 \text{ inches.}$$

As the float is attached to a wire rope the other extremity of which wraps round the sector (t) of 28 inches radius, this deviation corresponds to 7′ 51″ of space at the equator in the first case, but only 3″·2 in the second case.

Assuming that the flow of the water from the box will be uniform if the head be uniform, the only remaining cause of irregularity lies in this, that the overflow is kept at a constant depth below any fixed point in the float ; that point, through variation of the resistance, not being at an absolutely invariable distance below the surface of the water in the box ; consequently, the head will vary if the immersion of the float varies.

We have seen above that, with a variation of friction of 3 lbs., the variation of immersion $= ·064$ of an inch ; therefore (47 inches being the distance of overflow below the surface of water) the ratio of maximum velocity to mean $= \sqrt{\dfrac{47·064}{47}} = \dfrac{1·00067}{1}$, which corresponds to an angular deviation in right ascension of 36″ of space at the equator per hour.

In practice, however, when the instrument is accurately counterpoised and carefully oiled, the variation of the force required to turn it round in right ascension is probably very much less than this ; and if between one night and the next the friction increases through viscosity of the oil, the rate is easily corrected by means of the valve (o).

This clock has been used for micrometrical observations, and answers perfectly ; but how far for spectroscope purposes it will bear comparison with other contrivances, has not yet been ascertained.

The accompanying figure has been drawn on a scale of ⅓ inch to 1 foot, with the exception of the sector (t) and the weight (s), which keeps the rope (x) extended ; these two parts having been drawn on a reduced scale, and in a different position to save space.

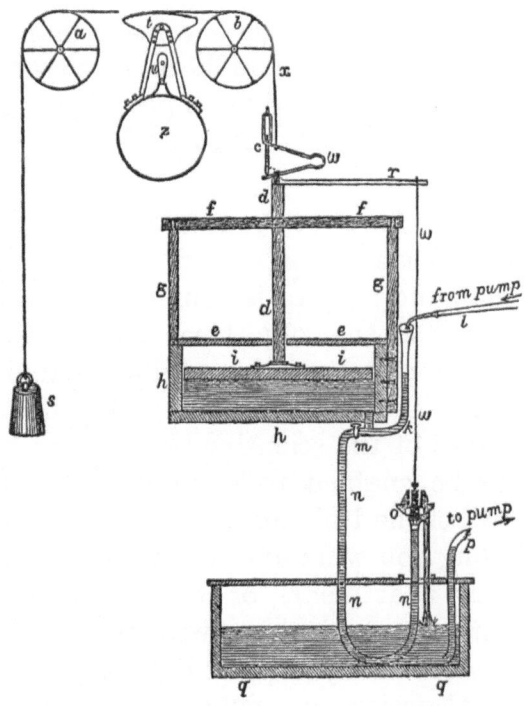

THE ROYAL SOCIETY

Address of the Rt. Hon. the Earl of Rosse,
The President,
Delivered at the Anniversary Meeting of the Royal Society,
on Thursday, November 30, 1854

GENTLEMEN,

WHEN we met last November, I ventured to remark that the objects of science were better understood than they had been formerly, and that accurate notions on scientific subjects were becoming more prevalent ; the progress in that direction has still continued, and during the last year it has been even more decided than before.

With the spread of knowledge, the great body of the community are now better enabled to appreciate the importance of science in promoting the moral and physical improvement of mankind ; there is consequently a growing desire that science should be advanced, and as that object has been best effected by scientific societies, they are regarded with more interest. Public opinion having taken that direction decidedly, it was not unreasonable to expect that Government would feel anxious to meet the wishes of the scientific bodies, and provide them with a building where they would be enabled to employ to the best advantage the machinery of association, which had already effected so much. It is probable, therefore, that before very long you will have the important question to decide— whether to retain your present apartments, or to accept a suitable place in a new building. Throughout all the communications with the Government your officers have taken especial care to be perfectly explicit on two points : one, that till the plans were prepared, any final decision on your part was impossible ; the other, that you had a free option to retain your present apartments should you think fit, having received them as a grant, to endure so long as the Society should exist. There can now, I think, be little doubt but that the leading Scientific Societies will be suitably provided for, and that you will have the opportunity of taking your proper place at the head of science, with the other Societies by your side.

The pursuit of science will thus be greatly facilitated ; it will be rendered more convenient, and therefore more attractive ; and many will devote themselves to it who perhaps otherwise would not have entered upon it at all.

While, however, the Government is wisely anxious to encourage the active pursuit of science by meeting the wishes of the great body of scientific men, we perhaps may have it in our power to effect something in the same direction.

From various observations of my late lamented predecessor in his Addresses, it was evidently his opinion that we should act wisely in shaping our rules so as to adapt them to the varying usages of society, rather than to preserve everything unchanged, relying on the sanction of a long prescription. In former times, November was the height of the London season, and the Anniversary, with all its preliminary business, was naturally held then, because the largest number of Fellows were in town. Whether for the better or not, the season has been changed, and it is now six months later ; we, however, remain where we were. The Council meets the latter end of October, and continues its meetings through November ; the principal business being to award the medals, and to select the Officers and Council to be recommended to the Society at large for election. The November business is wound up by the Anniversary, the most important meeting of the Society. The Fellows who are not permanent residents in London are naturally absent in the country ; the great business of the year with all its responsibilities, devolves therefore upon a section of the Society. A country gentleman, if named upon the Council, cannot conveniently attend in November, and but few country gentlemen attend the anniversary.[a] Our proceedings are therefore not of as much interest to the Fellows taken as a whole as they might be made to be, and they are not calculated to attract the men who, having attained high scientific distinction at the Universities, reside in the country, unharassed by professional calls, and who have therefore both training and leisure, important elements of success in all scientific pursuits.

The award of the medals at a time when so many Fellows are absent is also attended with inconvenience ; for although an attempt is made to secure for each science a kind of representation in the Council, still so wide is the range of science now, that special departments are often necessarily unrepresented. In disposing of papers, the imperfect representation of individual sciences is unattended with inconvenience, because each paper is referred to two Fellows to report upon ; the Council thus calls the whole Society to its aid, and the result, I believe, is perfectly satisfactory. With the medals it is otherwise ; no official reference is made to Fellows not on the Council. There is a further difficulty. The questions which usually arise are of this nature. Discoveries have been made by different individuals in various sciences. Who has added most to the general stock of knowledge by a positive contribution ? Who has the merit of having effected discoveries of most promise ? Recollect, that in answering these questions some estimate must be made of the weight due to each science, for they cannot be considered all alike ; very far from it. Some sciences require great mental labour, guided by faculties of a very high order—a rare gift ; while other sciences can be cultivated successfully by common-place men, with only a moderate amount of perseverance. That such an estimate can be made, and which carries with it a kind of general assent, is evidenced by the fact that it is annually made, to some extent, at the examination for the Fellowship in the University of Dublin, which bears a certain loose but not uninstructive analogy to the process by which your Council awards the medals. That examination is the most difficult which exists ; the prize is large, and the competition is free. There is a course of mathematics comprehending, I may say, everything ; a course of physics equally large ; a very large course of moral philosophy ; a course of metaphysics, of logic, classics, history, chronology, and Hebrew. The examination is, of course, public ; and a person of experience, acquainted with the course, can usually at the close of the examination point out the successful candidate. Some have answered better in one science, some in another, but acting under the guidance of a mature judgment a kind of equitable adjustment has been made by the bystander, which has led him to the same conclusion as the examiners. Now, let us see for a moment how this has been brought about. The examiners, who are Fellows, are conversant to a certain extent with all the sciences ; and in measuring the value of each answer, they are governed by a well-marked public opinion in the University, precisely as is the case with the enlightened audience ; and they come to the same conclusion. But with your Council the case is necessarily very different. However chosen, they cannot have within themselves the same means of discharging their very difficult duties in a way which will carry with it the full concurrence of the Society. Take as an example the simplest case which can arise : two persons have been proposed for the medal, a chemist and a mathematician. Upon the Council we will presume there is a first-rate chemist and a first-rate mathematician. Now, in chemistry and in mathematics, and indeed in all the sciences, little discoveries are very abundant. By what possible means can the chemist bring his mind to bear upon the little discoveries of the mathematician, so as to weigh them, even in the roughest manner, against the discoveries in his own science ? Or will the mathematician be more fortunate in dealing with the discoveries of the chemist ? But how is it with respect to the other members of the Council ? There will probably be gentlemen representing the different branches of the natural sciences, also perhaps a geologist, an astronomer, an engineer. Why, even in the very simple case I have supposed, the elements for the roughest approximation to a true conclusion are not within the Council, and it cannot be otherwise. The Council must therefore travel beyond its limits for the necessary information. Individual members will naturally therefore, in an unofficial way, consult such Fellows not on the Council as are known to be conversant with the particular question at issue ; and, practically, if time permitted and opportunity offered, a large proportion of the ablest Fellows would exercise a guiding influence on the Council, leading them to correct conclusions. Information would even be sought without the Society ; the prevalent opinions at the Universities would be looked for, and the state of public opinion abroad in scientific circles would in many cases have great weight. I need hardly say, that as things are at present, this inquiry is impossible except to a very limited extent, and public opinion, even in our own body, can afford but little of that aid to the Council which it would do under more favourable circumstances.

If the medals were awarded in June, after discussion in several successive Councils at considerable intervals, while the great body of Fellows, the leading members of the Universities, and the foreigners who visit London were in town, each member of the Council would have immediate access to the best sources of information. Recently the experiment has been tried of proposing candidates for the medals before the recess, but without, I think, any practical advantage. Where the candidates are numerous, inquiries would be endless ; and it is only when the number has been reduced, when the doubtful questions have been put prominently forward by discussion, and a decision is imminent, that inquiries will be prosecuted with energy, and can be made with effect.

Finally, experience has shown that even in the transaction of the business of the nation, there is so much inconvenience in running counter to the habits and usages of society, that it is only in a case of necessity that Parliament is assembled in November.

With these or similar views, the subject was brought before your Council in 1845 ; and, as was announced by Lord Northampton, it was resolved to change the day of the anniversary to a season more generally convenient. In his Address the year after he states that doubts had arisen as to the legality of the change without a new charter, and no serious effort appears to have been subsequently made to surmount the difficulty.

I believe it is well known that Sir Humphry Davy's opinions on this subject were very similar to Lord Northampton's, and I have heard that he had deliberately committed them to paper, setting out fully his views as to the prospects of science in this country, and the position the Royal Society should hold. Such a document would be of great value, and I have anxiously inquired for it, but in vain.

This was the state of things at the time of my election as President, and after full consideration I took the first opportunity, at the anniversary in 1849, of expressing my entire concurrence in the views of Lord Northampton. It appeared to me, however, to be quite evident that there was not a strong and universal desire to change the day of the anniversary ; and that was to be expected. There are certain associations, hallowed by time, to which we all recur with pleasure : when we meet on the 30th of November, our thoughts are led back to the auspicious day when the Royal Society was founded : we are reminded of Boyle, Wren, Hooke and Wallis, the first Fellows, and feel a just pride that we enjoy the high honour of being their successors. Few, perhaps, would assent without some degree of reluctance to a change which would sever these ancient associations. Feeling assured, therefore, that there were various shades of opinion, and having stated my views distinctly in my first address, I did not conceive it to be my duty to proceed further. The next step would have been to have directed the attention of the Council to its former decision as to the expediency of changing the anniversary, and to have obtained the best advice as to the means to be taken to effect that object. I thought it better to let the matter rest for a time. Had I thought otherwise, there was certainly no one by whom such a question could have been brought forward with less propriety, or less advantage, than your President. He could not have recommended his arguments as springing from an unbiassed mind, when he was in the position, not a very agreeable one, of being necessarily absent during the autumn and winter, when there was so much business of importance. Now the case is different. There is no longer any reason for reserve, and in expressing in my last address the same opinions as in my first address, I have ventured to express them more strongly, because experience has more fully confirmed them. If we are to distribute medals, it is surely important that the award of the medals, like the award of the Fellowships in the University of Dublin, should carry with it the full approbation of the whole Society. If we are to have meetings to transact business, it is of the highest importance to make them easily accessible to all Fellows by the choice of a suitable season, so that every person, whether on the Council or not, might have the opportunity of exercising his privileges with the least amount of personal inconvenience. It is in this way we shall render the Royal Society even more popular than it is, and hasten its growth in strength and influence, so that it will become, not the Royal Society of London merely, but the Royal Society of the whole kingdom.

I have thus ventured to suggest certain changes as to the time and manner of transacting the business of the Royal Society ; the objects we have in view would, I think, be much promoted by another innovation, which it requires some courage to propose. I mean a large increase in the number of the Council. From what I have observed, I am convinced that to enable the Council to exercise an effectual supervision over all the sciences, it is necessary to make ample room, so that each science should be fully represented. There is another object perhaps of even greater importance to be attained by the same means, an *effectual representation of all classes* upon the Council, so that men of general attainments should have their place, and the government of the Society should not be exclusively entrusted to men, who, however eminent in especial branches of science, may not be always the most conversant with worldly affairs, or the most competent to transact that commonplace business, upon which, in the main, the prosperity of Societies depends, nothing would do more than such a change, to promote harmony and good feeling within our walls ; nothing would contribute more to increase the influence of the Royal Society in advancing the general interests of science.

I have already ventured to say that I had little doubt but that the memorial with its two hundred signatures in favour of juxtaposition, backed up by public opinion, would produce the desired result, and that before long we should see the leading scientific Societies under the same roof. I think I cannot now err in expressing a confident belief, that whatever changes may be required in our Society to meet the just wishes of scientific men, will be carried out with readiness. The progress of science will thus be promoted, and this country will gradually attain even a higher place in European science than that which it at present holds. There is a new quarter, however, to which science may I think look hopefully, and that is the University of Oxford. Last session an act was passed for effecting certain improvements in the University of Oxford. Under that act a commission was appointed, consisting of men of learning and high station, to advise and co-operate with the governing body, and so effect such changes as might be useful. At present it can scarcely be said that science at Oxford receives any substantial encouragement. The fellowships are for the most part close, and therefore are not necessarily the rewards of learning ; and where they are open, the success of the candidate depends upon his proficiency in the ancient languages and literature. Honours are indeed awarded to the mathematical and physical sciences, but they carry with them no emoluments ; and without any knowledge of the mathematical sciences except the elements of plane geometry, and without any knowledge whatever of the physical sciences, the highest University honours may be obtained. A man therefore after having very creditably passed a public school, and having taken his degree with a first class *in literis humanioribus*, may find that he knows no more than was known 1800 years ago. He may be ignorant of physics in its most elementary form, and may therefore be incapable of comprehending the first principles of machinery and manufactures, or of forming a just and enlarged conception of the resources of this great country. That the legislation of last session should long continue unfruitful, I think, is improbable, and the time seems to be at hand when the cultivation of the physical sciences will receive a new impulse at our universities, and when the great resources of Oxford will, in part, be applied as the rewards of scientific eminence.

You are all, Gentlemen, no doubt aware, that in 1823 your Council, at the request of the Lords of the Treasury, appointed a Committee to report upon Mr. Babbage's plan for the construction of a Calculating Machine, which he called a Difference Engine. The Committee, I need hardly say, was composed of men eminent for their theoretical and practical acquaintance with such subjects : that Committee recommended the Lords of the Treasury to assist Mr. Babbage in carrying out his undertaking. The Lords of the Treasury acquiesced, and the work was proceeded with ; Mr. Babbage exercising a constant and vigilant superintendence, furnishing the designs, making the computations, in fact supplying all the theoretical requirements, while the Government supplied the manual labour and raw materials. In the then backward state of mechanical engineering, great difficulties were encountered ; at length in 1828 the Royal Society was again consulted by Government, and the result was a report from a Committee, to the effect that satisfactory progress had been made, considering the difficulties, and that the engine was likely to answer the expectations of its inventor. The Council adpoted the report, and communicated it to Government, with a strong recommendation in favour of the undertaking. The Government acting under that recommendation supplied further funds, on the condition that the engine was to be public property, and the work proceeded. In 1830 the Royal Society was again consulted by Government, and the Council acting as on former occasions appointed a Committee.   The report which was drawn up in a detailed form was satisfactory to the Treasury, and the Council were informed that funds would be supplied from time to time till the engine was completed.   Very soon a new difficulty occurred ; it became necessary to change the engineer, and it was then found that by the rules of the trade, the tools, which had been constructed at the public expense, were the private property of the engineer ; there was no choice, therefore, but to sacrifice the tools, or to endeavour to effect a compromise for a large sum.   The progress of the work was suspended : there was a change of government.   Science was weighed against gold by a new standard, and it was resolved to proceed no further.   No enterprise could have had its beginning under more auspicious circumstances : the Government had taken the initiative, they had called for advice, and the adviser was the highest scientific authority in this country ;—your Council, guided by such men as Davy, Wollaston and Herschel. By your Council the undertaking was inaugurated, by your Council it was watched over in its progress.   That the first great effort to employ the powers of calculating mechanism, in aid of the human intellect, should have been suffered in this great country to expire fruitless, because there was no tangible evidence of immediate profit, as a British subject I deeply regret, and as a Fellow my regret is accompanied with feelings of bitter disappointment.   When a question has once been disposed of, succeeding Governments rarely reopen it, still I thought I should not be doing my duty if I did not take some opportunity of bringing the facts once more before Government.   Circumstances had changed, mechanical engineering had made much progress, the tools required and trained workmen were to be found in the workshops of the leading mechanists, the founder's art was so advanced that casting had been substituted for cutting in making the change wheels, even of screw-cutting engines, and therefore it was very probable that persons would be found willing to undertake to complete the Difference Engine for a specific sum.

That finished, the question would then have arisen, how far it was advisable to endeavour, by the same means, to turn to account the great labour which had been expended under the guidance of inventive powers the most original, controlled by mathematics of a very high order ; and which had been wholly devoted for so many years to the great task of carrying the powers of calculating machinery to its utmost limits.   Before I took any step, I wrote to several very eminent men of science inquiring whether in their opinion any great scientific object would be gained, if Mr. Babbage's views, as explained in Ménabréa's little essay, were completely realized.   The answers I received were strongly in the affirmative. As it was necessary the subject should be laid before Government in a form as practical as possible, I wrote to one of our most eminent mechanical engineers to inquire whether I should be safe in stating to Government that the expense of the Calculating Engine had been more than repaid in the improvements in mechanism directly referable to it : he replied, unquestionably.   Fortified by these opinions I submitted this proposition to Government :—that they should call upon the President of the Society of Civil Engineers to report whether it would be practicable to make a contract for the completion of Mr. Babbage's Difference

Engine, and if so, for what sum. This was in 1852, during the short administration of Lord Derby, and it led to no result. The time was unfortunate, a great political contest was impending, and before there was a lull in politics, so that the voice of Science could be heard, Lord Derby's government was at an end.

Although, in communicating with Lord Derby, I was not acting under the directions of your Council, still, as my object was to induce the Government to complete a work in which this Society had taken so great an interest, I conceived it to be my duty to lay the facts before you, as a basis to proceed upon, should it hereafter be considered expedient to renew the subject.

I have detailed to you regularly at each anniversary the proceedings of the Committee of Recommendations ; a Committee, as you are all probably aware, appointed to distribute the grant for scientific objects which was made to us by Government the first year of my Presidency, and has since been continued annually. On the present occasion I have only to say, that since the last anniversary numerous reports have been received, and I hope the new Council will consider it expedient to collect the facts brought out, and arrange them in the form of a paper to be laid before Parliament.

With respect to the design for the re-examination of the heavens in the southern hemisphere, originally suggested by the British Association, and subsequently matured by your Council, I have only to say, as I said the last anniversary, that it is in the hands of Government.

There is no other subject which seems to me to call for observation ; the report of your Treasurer will give all necessary information as to financial matters, and it only remains for me to express the deep sense I feel of the great services which have been rendered to Science by your Councils during the six years I have been officially connected with them. I am sure nothing could have exceeded their pains-taking industry, their complete devotion to your service. In their hands your interests were watched over with anxious care, they were in perfect safe-keeping ; and when I was unavoidably absent, as was too often the case, I had no misgivings. To your Council I return my sincere thanks ; to you, Gentlemen, I feel equally grateful ; and in retiring from your Presidency permit me to assure you, that although the position I am destined henceforth to occupy will be less prominent, my exertions for the welfare of your Society shall not be less earnest.

From the PHILOSOPHICAL TRANSACTIONS *of the*
Royal Society

XXII. *An Account of Experiments on the Reflecting Telescope. By the Right Honourable Lord* OXMANTOWN, *F.R.S., &c.*

Received May 9,—Read June 18, 1840.

FOR several years I have been engaged in a series of experiments, in the hope of increasing the power of the telescope, and I am induced, perhaps rather prematurely, to lay some account of them before the Royal Society, conceiving that, from the scale upon which they have been carried on, and perhaps from their results, they may prove interesting, particularly to those who have devoted their attention to such subjects. I should have been glad to have postponed this communication to a more distant time, so as to have rendered it in many respects more complete, but as experiments on a large scale are necessarily very tedious (months, even years, passing away almost imperceptibly), to have done so would have been to have postponed it indefinitely, and to have withheld facts which may perhaps be useful to those who are engaged in similar pursuits.

All the experiments I am about to describe relate solely to the Reflecting Telescope. With the exception of a few trifling experiments many years ago on fluid object-glasses, which led to no result, I have not had any experience in the construction of the refractor.

I have long thought that in the present state of knowledge there was not much prospect of improving the refractor to any considerable extent; the fluid object-glass, at least in my hands, did not appear promising, and the improvements which have been made in the manufacture of glass on the continent, seem to have effected little more than to afford the means of constructing larger discs of tolerably perfect glass, than was formerly practicable; still wanting, however, that exact homogeneity, and indeed those optical properties essential to any great increase of power. Upon the whole, therefore, there seemed to me to be but little chance of effecting anything really important in the present state of astronomy, except by improving the reflecting telescope; to that object, therefore, every effort was directed. The task was evidently a very difficult one, as the late Sir W HERSCHEL had apparently almost exhausted the subject, having devoted to it acquirements the most varied and extensive, and at the same time the most suitable, during a very long life. Still it did not seem impossible that, profiting by his labours, and imitating his example of steady perseverance, some advance might be made, trifling perhaps, but eventually leading to valuable results.

The great object seemed to be, to remove, as far as possible, the causes, owing to

Fig. 1

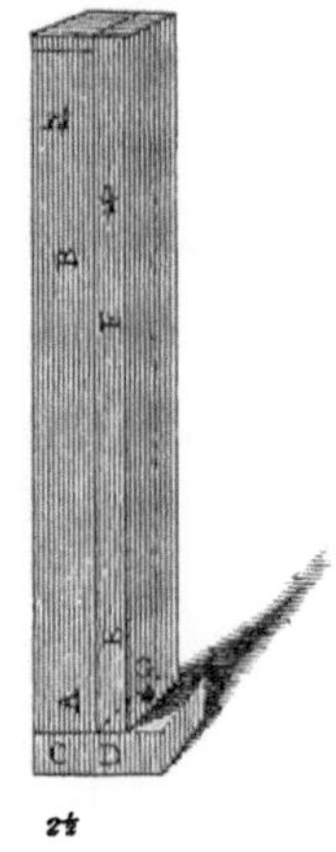

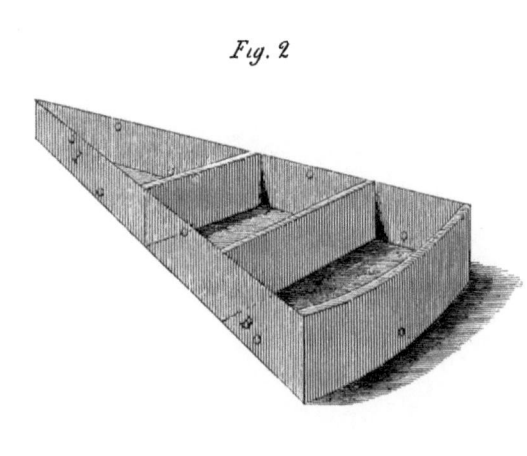

Fig. 2

Fig 3.

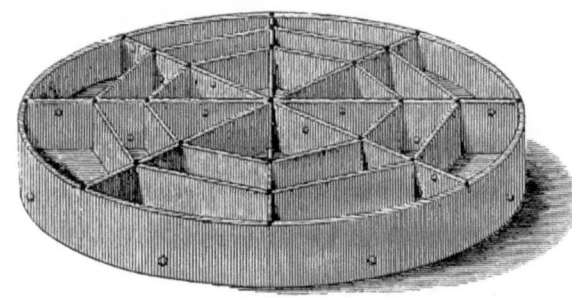

Fig 4.

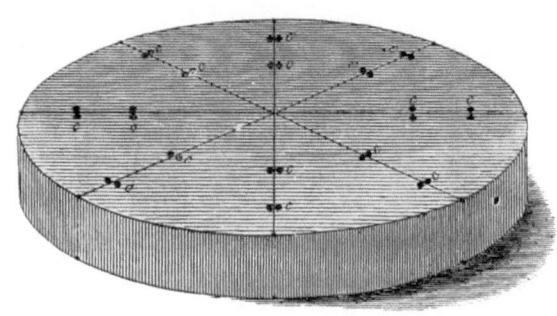

Fig. 5.

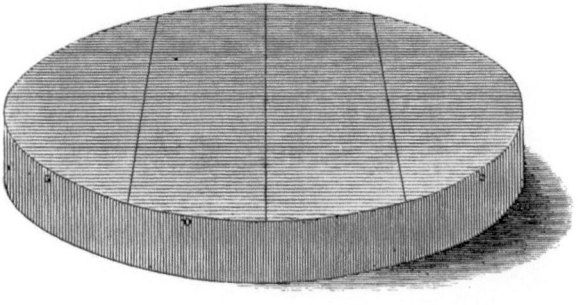

which, in increasing the size, much of the perfection of smaller instruments had invariably been sacrificed. For instance, to avoid the brittleness of the best speculum metal, it had been found necessary, as the dimensions were increased, to use an increased proportion of copper, so that the alloy was inferior in brilliancy, yellower, and much more liable to tarnish*. In polishing large surfaces also there were great and peculiar difficulties, all the defects having a tendency to augment rapidly with the size, and proportionately to impair the defining power. I thought, therefore, that were it possible to discover means of securing, in the construction of large instruments, the same excellence of material and accuracy of surface, which on a small scale was attainable by the processes already published, much would be gained; but even in the construction of small instruments there was no reason to suppose that the utmost limits of perfection had already been attained. On the contrary, the unequal performance of specula wrought with the utmost skill by the same hand, and with the same materials, was a fact which clearly proved that certainty of effect, the usual characteristic of a perfect mechanical operation, was still to be sought for. It was evident, therefore, that there was ample room for experiment with a prospect of some success. With these views, the first experiments were undertaken: they were on a small scale, the earliest as far back as 1827, and some account of them appeared at the time in Sir David Brewster's Journal for July 1828. Since that time, the scale was gradually increased, till an aperture of three feet was attained, beyond which as yet I have not proceeded.

I should, perhaps, however, remark here, that although these experiments were exceedingly numerous, they were not near so tedious as might be supposed from the length of time that has elapsed since their commencement; for in the midst of other avocations, long interruptions occurred, so that probably more might have been accomplished in one-third of the time employed continuously: there is nothing, therefore, really calculated to deter others, who have the means, from devoting their time to the improvement of the reflector: the joint labour of many may effect much, and the object is of no less importance than the advancement of those inquiries in practical astronomy, in which for further progress we must now look solely to an increase in the optical power of instruments.

In describing these experiments, it will be necessary to enter somewhat into detail, so as to enable others to repeat them without much difficulty; and as success often depends upon attention to minute particulars which have only been observed after repeated failures, it will be impossible to be as concise as it would otherwise be desirable; nothing, however, shall be inserted which has not been repeatedly tried with care, and is not calculated, either by its success or failure, to be practically useful: and first as to the best materials for constructing the reflecting surface. On this subject many experiments have been tried, but upon the whole I have little to add to what is generally known: tin and copper, the materials employed by Newton in the

* Pearson's Astronomy, p. 74; and article Telescope, Rees's Cyclopedia, Smeaton's Letter.

first reflecting telescope, are preferable to any other with which I am acquainted; the best proportions being four atoms of copper to one of tin (TURNER's numbers), in fact, 126·4 parts of copper to 58·9 of tin. This alloy, however, as well as every specimen of speculum metal I have ever examined, is visibly porous when carefully tested with a microscope; a very high power is not necessary: CODDINGTON's microscope, even with its lowest power, is generally sufficient; with his highest power always so. When the copper is very impure, and the alloy has been much heated, then the pores are often easily perceptible to the naked eye: contrary, however, to what might have been inferred from this, by making the copper and tin chemically pure, we do not get rid of them altogether; pores can still be detected by the microscope, even when the metal has been melted at the lowest temperature. Good cast steel is free from pores, but cannot be hardened without cracking when of considerable dimensions: all copper is, I believe, porous, tin not so.

NEWTON, with his usual acuteness, observed this porousness in speculum metal, and considered it a serious defect; and, although, where the process of polishing is conducted in the best manner, the evil which he had apprehended, viz. the rounding of the edges of the cavities, is not perceptible, still it is probable that it always takes place in some degree, however trifling, producing a defect so mixed up with other unavoidable errors as to remain undetected. The smallest defect, however, deserves attention, and perhaps others may succeed in discovering the cause of this, which I have not: my experiments lead me to think it is in some way connected with the presence of carbon or oxygen; and that possibly by watching the toughening of copper on a great scale, a process which chemistry has not satisfactorily explained, some information might be obtained.

Upon the whole, therefore, there is at present no sufficient reason for making use of copper in any other state than that in which it can be procured from the merchant, except perhaps for the small flat metals of the Newtonian; and with block-tin in the proper proportions, we have upon the whole the best material yet discovered for the specula of reflecting telescopes. It is, however, an alloy which has always been found rather difficult to manage; it is not easily cast sound, even of moderate dimensions, by the ordinary processes of the founder, or by those which have been published by scientific men; and even when so cast is very liable to break in cooling, or afterwards by exposure to any slight but sudden change of temperature, sometimes breaking even without any apparent cause.

The difficulties appeared so great, that reflecting upon SMEATON's account of Sir W. HERSCHEL's labours, I had very little hope that any very large specula could be cast and safely polished of so brittle and untractable a material; and although some of the difficulties have been surmounted, still it is probable, that in seeking to obtain the utmost possible amount of telescopic power the state of our atmosphere admits of, the requisite dimensions will not be attained by a simple casting of this alloy. The idea, therefore, obviously suggested itself of uniting several castings into one re-

Phil.Trans.MDCCCXL.Plate XXI.p 515

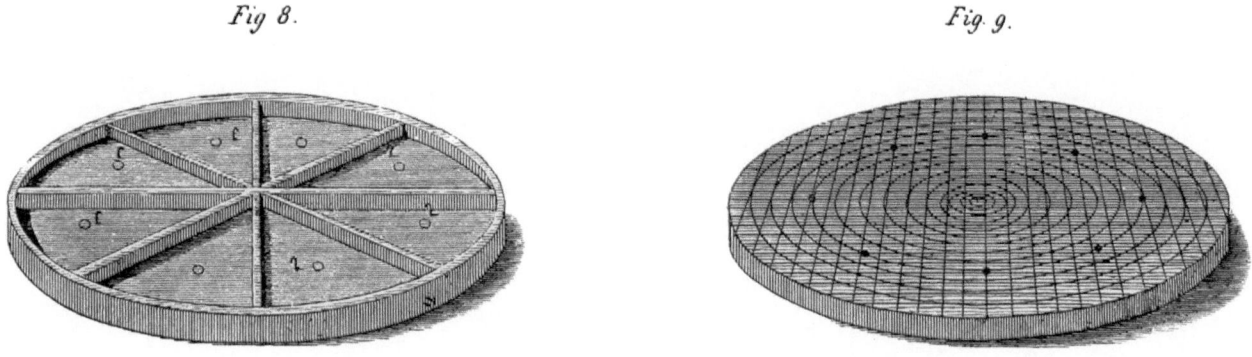

Fig. 6.

Scale of Feet

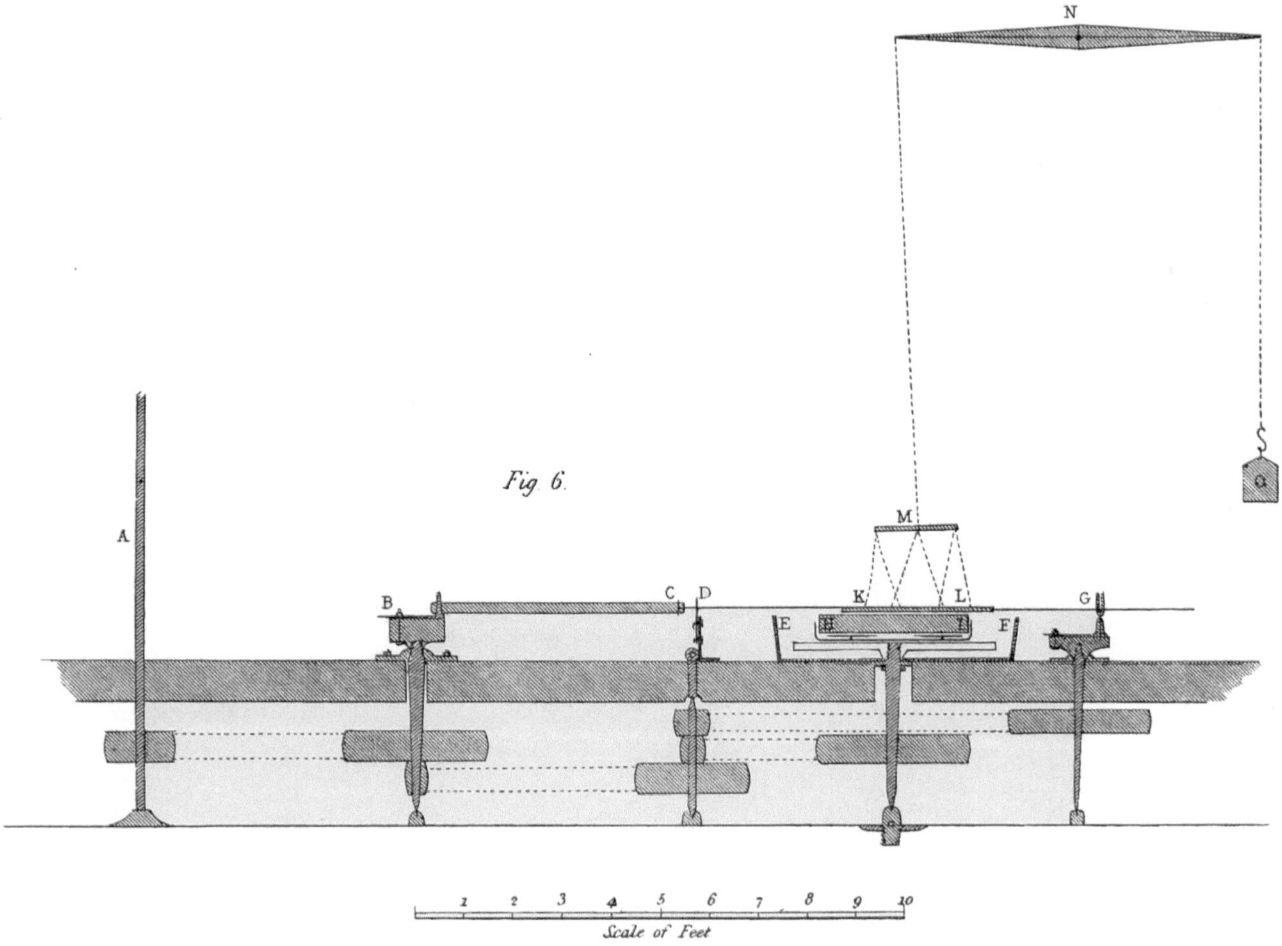

Fig. 8.

Fig. 9.

Face page 82

flecting surface, and as the best means of effecting that, to solder them upon an alloy of zinc and copper, (in fact, upon a species of brass,) which should expand and contract in the same proportion as speculum metal.

The experiment was tried repeatedly upon a small scale, and the difficulties which presented themselves successively surmounted, till at length a speculum of three feet aperture was completed, which bears distinctly a high power.

The evil most to be apprehended in this construction, was unequal contraction and expansion in the different changes of atmospheric temperature; that, however, seems to have been sufficiently guarded against, as I have never succeeded in detecting the slightest change of figure. It is therefore highly probable, that specula very much larger may be constructed upon this plan, and that, in fact, by repeating a number of operations, each on a scale perfectly manageable, we may obtain a reflecting surface the most brilliant, and at the same time as large as can ever be usefully employed in astronomical researches. I will now proceed with the details of the construction of the three-feet speculum. I had previously ascertained that an alloy consisting of about 2·75 of copper to 1 of zinc, gave the same contractions and expansions as speculum metal; it was necessary, however, to repeat the measurement with care, using the identical materials which were to be employed in the construction of the speculum. The following simple contrivance for obtaining the relative contractions and expansions of a bar of speculum metal, and different alloys of zinc and copper, had been employed in the previous experiments, and was again resorted to.

A bar was cast of speculum metal fifteen inches long, and one inch and a quarter square; similar bars, but only three-fourths of an inch thick, were cast of the alloys to be tried, containing a little more or less zinc than the proportions I have given. A piece of brass, consisting of 2·75 of copper to 1 of zinc, was also cast and soldered to the bar of speculum metal, as represented in Plate XX. figure 1, where A B is speculum metal, C D brass, and E F the bar of alloy, with a small excavation in the lower end fitted by grinding, so as to rest steadily on the hemispherical disc G; a thin slip of brass was also soldered at the upper end of the bar of speculum metal, and the two bars made to fit neatly there, so that when brought together, a very fine line could be drawn across them with scarcely any troublesome parallax at the joint; the whole was then immersed almost to the top in a tin vessel of water of the temperature of the atmosphere, and that vessel placed in another much larger, also containing water. Pieces of ice were then dropped into the outer vessel, so that the temperature of the whole was evenly and gradually brought down nearly to 32°; and a straight line, as fine as possible, was then drawn across both bars, and examined with a microscope to ascertain that it was perfect. The temperature was then gradually raised by pouring hot water into the outer vessel until nearly 212° had been attained, and the line was again examined with a microscope, and where the alloy had been made by mixing 2·74 of copper with 1 of zinc, and the loss in melting amounted to $\frac{1}{180}$th of the whole, the continuity of the line was not broken in that range of temperature; according,

G

however, as the proportion of the zinc was more or less, the expansion of the brass bar was greater or less than that of speculum metal.

As speculum metal and brass cannot be soldered together, except on a very small scale, with certainty by the ordinary methods practised in the arts, it may be useful to mention, that all that is necessary is, first, to fit the brass and speculum metal nicely, by filing or grinding, according to circumstances; the brass is then to be tinned, and suffered to cool; the surface of the speculum metal should be scraped lightly with a sharp chisel all over; the two surfaces are to be placed in contact, and matters so managed, that a slight pressure may be applied after the fusion of the tin, and continued until it has again become solid; the temperature should then be gradually raised till the tin melts; and then, but not till then, resin applied in fusion, and also a little melted tin; if resin or tin is applied in the solid state, owing to their rapid absorption of heat in becoming fluid, they will crack the speculum metal: the surfaces may be slightly separated, so as to ascertain that the speculum metal is tinned all over, which will be the case when the temperature reaches 400°: the whole must then be suffered to cool gradually.

In casting the alloy of zinc and copper, some precautions are also necessary, not practised and not required in the arts. At the commencement of these experiments, a difficulty occurred which for some time I failed in overcoming; in making the alloy of zinc and copper, and recasting it when made, much of the zinc was always volatilized, and the amount lost differed considerably each time, though the process was conducted as nearly as possible in a similar manner, so that it was impossible to allow for it, and the composition of the alloy was therefore uncertain; a fatal objection to the whole process. The furnace was the ordinary brass-founder's furnace, the air furnace of the chemist. I found, however, that the zinc was not volatilized in its metallic state; its affinity for the copper was sufficient to prevent that, but it was first oxydized. To prevent the fresh air from beating down upon the crucible whenever the lid of the furnace was removed, the furnace was made much deeper; this was of great use, but still the air which had passed through the fuel retained oxygen enough to act considerably on the zinc. Charcoal was next heaped over the crucible, still this was not effectual; at length, charcoal in fine powder was tried in a layer two inches thick on the surface of the metal, occasionally renewed; this was effectual. The crucible, therefore, must not be filled completely, and the charcoal dust can be conveniently thrown into it, whenever required, folded up in paper. If the process has been conducted with these precautions, the loss will be about $\frac{1}{180}$th, and almost exactly the same each casting.

The proportions of the zinc and copper having been determined, the brass work was first cast. By a reference to figures 2, 3, and 4, Plate XX., it will be at once seen in what way the materials were disposed of. Figure 2. represents one-eighth part of the whole seen in the reverse, a single casting; fig. 3. the whole speculum, also seen in the reverse; and fig. 4. the speculum seen on the opposite side previous to the soldering on of the plates of speculum metal; fig. 5. the speculum complete, faced with sixteen plates of speculum metal. The whole depth of the brass work was five inches and a half, and weight about four hundred and fifty pounds; the sides I B (fig. 2.) were made square and true, and then tinned, and the whole bolted together with iron bolts, as represented in the figure; the temperature was then raised till the tin was in fusion, and the bolts tightened to the utmost, so as to make close joints; in this state the speculum was turned, plated with speculum metal, and polished; but it was found that the joints had not been sufficiently secured, owing to which the bracing had not produced its proper effect, and flexure was quite perceptible by its effects in the different positions of the telescope; on a close examination, there was reason to suspect that the solder was not everywhere perfect, though it had been so originally; that in the rough operation of bolting to the lathe and turning, it had been in some places detached, and had not subsequently united, though it had necessarily been fused when the plates of speculum metal were soldered on: it was evident, therefore, that tin alone was not to be depended upon in putting together a speculum of this size. The iron bolts, of course, contracting less than the brass, had added nothing to the security of the joints; brass bolts had been tried formerly in constructing a speculum two feet diameter, but they were not strong enough to bring the joints close, and were replaced by iron ones.

The plates of speculum metal were, therefore, taken off, and the joints secured in the following manner. The whole mass of brass work was imbedded in casting sand everywhere in contact with it; over it there were about three inches of sand. The sand was then removed from the centre, so as to expose a circular surface of about one inch and a half in diameter, and about forty-five pounds of very hot melted brass, in the same proportions as the brass speculum, were poured upon it in a continued stream from a perpendicular height of about ten inches: the sand having been so arranged, that after the melted brass had reached the depth of two inches, the remainder continued to flow off; the melted brass as it began to cool was therefore continually replaced with hot brass; and from the height at which it was poured, it was necessarily in immediate contact with the surface of the cold metal, and soon completely fused it, and perfectly united it in that place. This was repeated in thirty-four different places, marked by dots in the figures: it was also tried in the places marked c, c, (fig. 3.) but failed. When the redundant brass had been removed with a saw, and the surface made smooth, a slight hair crack was perceptible at the boundary of the fused metal, proceeding no doubt from expansion and contraction after the brass had ceased to be fluid, but before it had become ductile. This process is made use of by the brass-founder in stopping holes in defective castings; he calls it *burning*. The failure of this process at the places marked c, c, is a fact which it would be important to keep in view in disposing of the parts of a large speculum. As it was desirable that some further attempt should be made to secure the joint at c, c, holes of about one inch and a half in diameter were bored completely through at c, and the intermediate brass chiseled out, as represented in the figure. They were then filled with melted brass, which was easily effected by imbedding the whole in sand; clearing the sand completely out of the spaces, and pouring the brass into each through a hole in a flat surface of sand packed in a small flask; when the brass cooled, the contraction was sufficient to draw the joint firmly together. This process may, possibly in engineering, prove useful as a means of connecting large masses of metal: I am not aware of its having before been put in practice. Conceiving that any further yielding in the joints was now impossible, the speculum was again placed on the lathe and turned to a radius of fifty-four feet, and new plates of speculum metal prepared for it.

Notwithstanding the small dimensions of these plates, none exceeding nine inches square, more difficulty was originally experienced in casting them than could have been anticipated. A great many unsuccessful attempts were made to cast them in sand, according to the directions of Mr. EDWARDS and others; they were seldom free from flaws, and, although cooled very slowly in an oven, they were extremely brittle, sometimes flying in pieces the moment they were touched, and generally breaking in the attempt to heat them again for soldering. The cause of this brittleness was evident, as the broken parts when replaced no longer fitted exactly, the metal therefore had been in a state of tension, owing no doubt to the edges of the plates becoming solid sooner than the centre.

The next plan tried was this: a number of equidistant thin plates of iron were immersed in a square crucible of cast iron filled with fluid speculum metal, so as to divide the whole mass into plates of equal thickness: this failed altogether, the plates of speculum metal were full of flaws; their contraction had been prevented by the unavoidable irregularities of the plates of iron, and crucible. Another plan was then tried, more successful than any of the preceding. A circular sawing machine was constructed, the blades were of soft iron, and while revolving were always partially immersed in emery and water; with this a block of speculum metal was without much difficulty cut into plates, which were perfectly free from flaws, and not liable to crack. Still their texture was not uniform; for about three quarters of an inch from their edgés the arrangement of the particles was different from what it was in the remainder; so much so, that the edge of the plates for that distance evidently resisted the action of emery more than the remainder, and therefore probably a speculum made of such plates would not eventually be as true as if they were free from that defect. Upon this point, however, I cannot speak quite decisively; from what I have observed I have strong reasons to think it is so; but were it of any importance to determine it, further experiments would be necessary. I have a speculum two feet aperture formed of such plates; it was however polished long ago, and I have not a polishing tool of the construction I now make use of, of the proper size; so it has not had a fair trial. The same plates were also originally used for the three-feet speculum I am describing, but when it became necessary further to secure the joints, they were taken off and spoiled, and in the meantime another process suggested itself, which seemed to be decidedly preferable.

It was evident that the flaws of so frequent occurrence in the plates formerly cast, and also their extreme brittleness, arose from the contraction of the metal in some places more than others, just at the time of transition from the fluid to the solid state. The edge of the plates always became solid first, and the central portions thus prevented from contracting were strained when no longer ductile. Were it possible, therefore, to satisfy the following conditions, viz. that heat should be abstracted rapidly and equally from the lower surface of a fluid disc of speculum metal, so that it should solidify from the bottom upwards in strata, or rather infinitely thin laminæ, the surface being the last to solidify, we should have a perfect casting; for the particles in that case being deposited, not uniformly, indeed, owing to the unknown action of the forces of crystallization, but in such a way as to fill up the interstices, there would be no flaws; and the température being uniform in a horizontal direction, and in the vertical varying in regular gradation from the lower surface to the upper, there would be no strain.

This, I believe, is the true principle upon which the most perfect castings can be obtained: its truth has been fully proved by practice; and, although in the arts fortunately it is not necessary often to attend to such minutiæ, as the materials employed are in some measure ductile, and therefore adapt themselves to the unequal strain to which they may be subjected, still it seems probable that the not unfrequent failure of large castings under a pressure much less than they were calculated to bear is due to this cause. The management of speculum metal may be regarded as an extreme case, where all the defects of manipulation are strikingly developed.

There are evidently two ways in which it might be possible to attain the required adjustment of temperature; the one by cooling the lower surface of a mould containing the liquid speculum metal, while the heat of the upper remained undiminished; the other by constructing the mould itself, so that the lower surface should absorb the heat rapidly and the upper retain it. Both were tried; the first by making the mould itself of cast iron, in which the metal was fused, and then exposing its lower surface to the action of a jet of cold water; the result justified the theory, but the mould very frequently cracked, and where this occurred before the speculum metal had become perfectly solid, the casting was spoiled; discs were, however, obtained of different sizes, the largest eighteen inches diameter, which was merely by chance, as a mould of that size almost invariably cracks before the completion of the process. The experiment, therefore, is not worth repeating, particularly as the other plan is simple and succeeds perfectly. It is obviously to make the lower surface of the mould of iron, and the upper sides of sand; at first a simple disc of iron was tried, but although the castings were sound, there was this defect; that bubbles of air were often entangled between the iron and speculum metal, producing cavities which it was tedious to grind out: the iron disc was therefore replaced by one made of pieces

G—1

of hoop-iron placed side by side, with their edges up, tightly packed in an iron frame: the edges were brought to a smooth surface of the proper curve either by the file or lathe, whichever was the most convenient. A metallic surface was thus constructed everywhere porous; as however close the hoop-iron had been packed, the interstices suffered air to pass freely through. So successful was this expedient, that of sixteen plates cast for the three-feet speculum, not one was defective: the following particulars require to be attended to. The disc of hoop-iron should be as thick as the speculum to be cast upon it, so as to cool it with sufficient rapidity; it requires to be warm, so that there may be no moisture deposited upon it from the sand; it may be heated to 212° without materially lessening its cooling power. The metal should enter the mould by the side, as is usual in iron founding, but much quicker, almost instantaneously; one second is sufficient for filling the mould of a nine-inch plate or speculum. As to the temperature of the metal, this can best be ascertained by stirring it with a wooden pole occasionally after it has become perfectly fluid; when the carbon of the pole reduces the oxide on the surface of the metal, rendering it brilliant like quicksilver, the heat is sufficient. When the metal has become solid in the ingate or hole through which it enters the mould, the plate is to be removed quickly to an oven heated a little below redness, to remain till cold, which, where the plates are nine inches diameter, should be three or four days at least.

The metal which had filled the ingates having been separated from the plates by a file, they were fitted to the brass speculum by grinding each separately by hand upon it till brought to the same curve. The surface of the brass speculum was then scraped and tinned, and when cold all resin was removed by washing it first with spirits of turpentine, and then soap and water. The lower surfaces of the plates which fitted the brass speculum, were carefully scraped all over with broad flat chisels of thin steel, perfectly hard; it is necessary that no spot should be passed over unscraped, and that the plate should be kept perfectly dry till soldered to the speculum. The plates were then arranged in their places on the brass speculum, which had been previously placed in an oven so constructed that the temperature could be gradually and equally raised. The bottom of the oven consisted of a cast-iron plate four feet diameter, set in brick work, as if it was intended for a sand bath; the brass speculum rested upon this, with loose bricks between to protect it from radiation, so that it should acquire heat solely from the heated air; the top of the oven had a cover in four pieces, that one might be opened at a time. In about eight hours the tin on the speculum was fused, and then melted resin was poured in between the plates; tin in fusion was also applied in the same manner, and the plates moved a little backwards and forwards. As soon as from the aspect of the edges of the plates it was certain that the tin was acting on the speculum metal, the fire was almost all withdrawn, and the temperature was not suffered to rise higher. The joints of the plates were now made straight, and kept open about one-twentieth of an inch by chips of wood; the whole was then suffered to cool gradually, and in five days it was ready to be ground.

There are here also some points which it is necessary to attend to. The perfect union of the plates with the brass speculum depends upon the fact, that if a plate of speculum metal, scraped as directed, is laid upon a clean surface of tinned brass, and the temperature raised a little beyond the fusing point of tin, and then melted resin and tin applied, the plate of speculum metal will be immediately tinned all over, and the union of course will be perfect. If resin, however, is applied at the beginning of the process, and therefore exposed for hours to an increasing temperature, the resin, before the temperature has reached the proper degree, will have been decomposed, and will effectually prevent the success of the operation. In my earliest experiments, I was not aware of this fact, and was therefore obliged to turn the plates over repeatedly with a wooden tongs, to remove the decomposed resin; and, although where the plates were small this was practicable, in attempting to manage the plates of the three-feet speculum I failed: several were broken by unavoidable exposure to variation of temperature. I may perhaps as well mention, that formerly I tried the muriate of ammonia instead of resin, and also a variety of other processes, but none was completely successful but the one I have given. The resin should not only be in fusion, but where the plates are large it is more prudent to regulate the temperature by a thermometer. The same observation applies to the tin; I have found, however, that a portion of unmelted tin in the ladle, in contact with the fluid metal, was a sufficient guarantee against a too great disparity of temperature. The brick work should be perfectly dry, as a drop of condensed steam falling upon a plate would certainly crack it.

Before I proceed to the grinding and polishing of the speculum, I will conclude the remaining experiments on the process of casting. The ease and certainty with which perfect plates of speculum metal of moderate dimensions had been obtained, obviously suggested a trial of the same principles on a larger scale. A perfect disc of fine speculum metal, twenty inches diameter, was immediately obtained, and recently so large a disc as three feet was cast perfect the first attempt, and has been finished without accident.

Although the principles which guided the manipulation have been fully explained, there are some practical details necessary to success, which might not perhaps suggest themselves immediately to others who have not had the same experience in managing this intractable alloy that I have had; a few remarks may therefore be useful. The disc, when cast, was about three inches and three quarters thick, and weighed about thirteen hundred weight; the metal for it was fused in two cast-iron crucibles. In my earliest experiments, in consequence of the failure of a very large crucible of cast iron, one inch and three quarters thick, made for me at one of the principal London foundries, I concluded too hastily that cast iron would not answer. The first time the crucible was tried, after it had been about four hours in the fire, the speculum metal oozed through it; when cold the defective portion was bored out and stopped with a screw; the next time the metal oozed out in several places. The crucible was then broken, and it was found that the speculum metal had penetrated into the iron in various directions, and a small portion of speculum metal, when analysed, yielded iron. The reverberatory furnace was therefore resorted to, but I found that, owing to the rapid oxidation of the tin, the quality of the alloy was continually changing, and the result therefore was uncertain. In the mean time I found that the affinity of speculum metal for cast iron was very slight, although for wrought iron it was considerable ; that the iron detected in the speculum metal was owing to the corrosion of the wrought-iron screw, with which the hole had been stopped, and that the failure of the crucible had arisen, not from the unfitness of the material, but from defective workmanship. After some experiments in my own laboratory, I found that crucibles cast in the usual way, with the mouth down, were generally defective, the speculum metal passing through minute pores not visible externally, as quicksilver through the sap vessels of wood. When the crucible, however, was cast with the mouth up, there was no such defect ; we therefore have in cast iron a material capable of being formed into crucibles of unlimited dimensions. I cast one which held fifteen hundred weight, in which the speculum metal was made, but two crucibles of half the size were manageable with less machinery, and therefore were used in casting the speculum ; they were raised from the furnaces by proper tackling, and placed in iron swing frames, so contrived that each crucible of metal could be thrown almost instantaneously into the mould. The fuel I make use of is wood or peat ; the latter when of proper quality is preferable, as it yields a steadier fire, without the intense heat of coke, which would without great care endanger the crucibles. The mould was made in the same manner as in casting the plates, differing from it merely in shape and dimensions ; it was what founders technically call an open one. The disc of hoop-iron was made circular three feet six inches diameter, three inches and a half thick, and turned upon a lathe convex to a radius of fifty-four feet. The speculum was cast with a groove round the edge, so that it might be securely embraced by a circular clamp tightened upon it with screw-bolts, to which the proper tackling could be hooked whenever it was necessary to move it.

When the metal had become solid, but was still red-hot, a strong hoop, somewhat larger than the diameter of the speculum, was placed upon it ; to this hoop a chain from a windlass, passing through the annealing oven, had been previously attached, and by the action of the windlass the speculum was drawn into the hot oven, and every opening closed ; in about a fortnight it was cool, and was found to be free from blemish.

It is of course impossible to ascertain à priori, whether it would be practicable to obtain still larger discs of fine speculum metal by this process, and polish them without accident ; possibly it might, as the principal cause of fracture, unequal contraction, no longer exists. The question, therefore, will arise, whether in endeavouring to obtain still larger specula, to approach nearer that limit beyond which it is not permitted to us to pass, the better course would be to attempt, with the risk of failure, a single casting of very great size, or at once to have recourse to the expedient of combining together a number of small castings, a process perhaps more tedious, possibly less perfect, but more certain. However, a further comparison of the two specula as to defining power, under all circumstances, and with the utmost care, will determine the future course of these experiments; at present there is no appreciable difference referable to their very different principles of construction; they both are free from flexure in the different positions of the instrument, and have defined equally well when polished with equal success. With a single lens of one quarter of an inch focus, giving a power of about thirteen hundred, they have both shown satisfactorily the dots on the dial-plate of a watch more distinctly than a very good refractor with a much lower power.

Hitherto the processes I have described were so effectual in producing the desired result, that there seems to be but little room for improvement, except in the discovery of new and better materials, an event by no means probable; but the case is far otherwise in the remaining operation, that of polishing the speculum; there, though the experiments have been even more numerous, much still remains to be accomplished.

Before the speculum is polished, it is worked to a spherical figure by a process technically called grinding, where the mutual attrition of the speculum, and a mass of nearly equal size of some hard substance, eventually produces a figure nearly spherical; and that, notwithstanding the irregularities, however great, of the surfaces of either, or both, at the commencement of the operation. Several ingenious devices have been, indeed, from time to time, suggested, more or less independent of the process of grinding, among which, perhaps, the most remarkable is that of Mr. BARTON, who proposed to communicate the figure and the polish at once, by turning the speculum with a diamond, constrained, by very delicate machinery, to move in the proper path, and with a motion so slow that the resulting grooves should act on light as a polished surface; but when we recollect the extreme accuracy required, that an error of figure amounting to but a small fraction of a hair's breadth would destroy the action of a speculum, it is scarcely to be expected that any process can succeed in practice, which has not, like that of grinding, a decided tendency to correct its own defects, and to produce a result in which the errors may be said to be infinitely small in comparison with the errors in any of the previous steps from which it was derived. I need, perhaps, therefore hardly say that all my experiments have been directed to the one object, that of endeavouring to improve the common published process of grinding and polishing, particularly in its application to large surfaces; for although the accuracy usually attained is so great that we fail in detecting by mechanical means, among a variety of specula made at different times and by different persons, any deviation from the proper figure, still by optical means, in fact, by trial in a telescope, the defects are at once apparent; and we shall probably find among them examples of every grade of defining power, from the speculum which is almost perfect, to that which does not define at all. This difference, so great, is mainly owing to a variety of minute circumstances incidental to the process of polishing, which influence the result in a greater or less degree; among which, the most important are, variations in the extent or relative velocities of the motions by which the necessary friction is produced; alterations of temperature; or some accidental pressure during the process. With a view of obviating these causes of uncertainty at a very early stage of these experiments, a machine was constructed for grinding and polishing, where the different motions were susceptible of separate adjustment, and were all under complete control. The first trials with it were upon the whole satisfactory, and I sent a sketch of it to Sir D. BREWSTER's Journal for October, 1828, in the hope of directing the attention of practical men to the subject, and perhaps of raising a doubt in their minds as to the justness of a very deep-rooted opinion, that specula could not be polished successfully except by the hand; an opinion which, if unfounded, must necessarily have been a serious obstacle to improvement, by precluding the use of the only means available in making accurate experiments under circumstances identically the same, or, indeed, of trying any series of experiments on a large scale. The machine was soon after enlarged, so as to be capable of working a speculum three feet diameter as its maximum, and otherwise improved, and since that no further alteration has been found necessary. From an experience of several years, during which specula have been ground and polished with it many hundred times, I can safely say that it fully answers the purpose, and I believe, in working large surfaces, a degree of precision can, with certainty, be obtained by it, unattainable by the hand, even by accident. The machine, in its present state, is represented in Plate XX. fig. 6, and Plate XXI. where A is a shaft connected with a steam-engine; B an eccentric, adjustable by a screw-bolt to give any length of stroke from 0 to 18 inches; C a joint; D a guide; E F a cistern for water, in which the speculum revolves; G another eccentric, adjustable like the first to any length of stroke from 0 to 18 inches. The bar D G passes through a slit, and therefore the pin at G necessarily turns on its axis in the same time as the eccentric. H I is the speculum in its box immersed in water to within one inch of its surface, and K L the polisher, which is of cast iron, and weighs about two and a half hundred weight. M is a round disc of wood connected with the polisher by strings hooked to it in six places, each two-thirds of the radius from the centre. At M there is a swivel and hook, to which a rope is attached, connecting the whole with the lever N, so that the polisher presses upon the speculum with a force equal to the difference between its own weight and that of the counterpoise O. For a speculum three feet diameter I make the counterpoise ten pounds lighter than the polisher. The bar D G fits the polisher nicely, but without tightness, so that the polisher turns freely round, usually about once for every fifteen or twenty revolutions of the speculum, and it is prevented by four guards from accidentally touching the speculum, and from pressing upon the polisher by the two guides through which its extremities pass. In fig. 7. this bar is on a larger scale. I have tried a variety of contrivances for connecting the machinery with the polisher, but the one I have described is by far the best. The wheel B makes, when polishing a three-feet speculum, sixteen revolutions in a minute; to polish a smaller speculum the velocity is increased by changing the pulley on the shaft A. The machine is in a room at the bottom of a high tower, and doors can be opened in the successive floors, so that a dial-plate of a watch placed perpendicularly over the speculum can be examined at any moment. The dial-plate is attached to a mast, so as to be much higher than the tower and about ninety feet from the speculum; and a small flat metal and eye-piece, with its proper adjustments, completes the arrangements for a Newtonian telescope. This simple contrivance has greatly facilitated the progress of these experiments. As appears in the plate, all the motions are produced by bands three inches wide, instead of the more permanent gearing of cog-wheels: although bands are liable to break, I think they are preferable to cog-wheels, because in a machine like this, which is for experimental purposes, if any part should become fast, which has happened more than once, the band falls off or breaks, and no mischief is done: it is only in a manufactory, where there are no experiments, but merely a routine of unvaried operations, that such accidents can be guarded against. One-horse power is quite sufficient to drive this machine while it is working a speculum of three feet diameter. The engine, however, is three-horse power, as from the distance of manufactories, it has been necessary to execute all the turning and casting work in my own laboratory.

The first serious difficulty which presented itself in polishing specula of considerable size, according to the common process, was this: when the layer of pitch was thin, as it must be to produce a good figure, however accurately at first it fitted the speculum, it soon ceased to do so, and the polishing did not of course proceed properly. This derangement, which in the ordinary mode of polishing by the hand is perceived at once by the feel, is as soon perceptible with the machine, because minute bubbles from the air, which has insinuated itself between the speculum and the polisher, are immediately observable. For some time this difficulty was exceedingly puzzling, and it was not until after many abortive attempts that the cause became evident: during the operation of polishing, the abraded matter, mixed with the polishing powder, is in part taken up by the pitch, but not equally over the whole surface; as, however, pitch is not sensibly elastic with a moderate pressure, wherever most is taken up, there the surface will be most prominent, and the figure of the polisher destroyed, unless the pitch can spread laterally. To allow of this lateral expansion a certain thickness of pitch is necessary, and I found, as might indeed have been anticipated, that the thickness required to be increased with the size of the speculum; in fact, if I may be allowed so to express myself, that the necessary thickness was some function of the diameter of the speculum. By using a layer of pitch sufficiently thick, a solid speculum of twenty inches aperture and a divided speculum of twenty-four inches aperture, the subjects of these and many other experiments, were made to define tolerably with a low power, and, at the same time, had acquired a high polish.

Mr. EDWARDS mentions that he had found, that, unless the pitch was of sufficient thickness, it would not preserve its figure; he had observed the fact correctly: in assigning the cause he was evidently in error, as he attributed it to the circumstance that the thin coat of pitch, as he supposed, acquired more heat from the friction than the thicker, while the reverse must have been the fact, as the thinner the imperfect conducting material, the quicker the metallic plate, to which it was attached, would have dissipated the heat, by radiation and conduction.

Mr. EDWARDS's specula were very small, and he found the thickness of half-a-crown sufficient. It was, however, evident, both as a deduction from all that had been written upon the subject, and from some experiments on small specula, that the necessity of using a layer of pitch thicker in proportion to the size of the speculum was a great evil, and was alone sufficient to make it impossible to polish large surfaces as accurately as small ones. A consideration of the theory, which I have ventured to put forward, suggested a very obvious remedy. It was very evident that by grooving the layer of pitch, provision might be made for its lateral expansion, wherever required, without so great a thickness. The experiment was first tried by reducing the thickness of pitch one-half, and making furrows in it by means of a hot iron quite down to, the metallic plate; the furrows were two inches apart, and there were two sets at right angles to each other, so that there were nowhere more than four square inches of pitchy surface in continuity. The result was, that the defining power of the speculum was immediately much improved. After many trials, however, a far better mode of effecting the same object suggested itself. The furrows were with difficulty kept everywhere open, and where there was a failure in this respect the old evil recurred, the polisher lost its figure; moreover, it was not found practicable to reduce the thickness of pitch to a minimum, which was a great object, and there were other minor practical inconveniences. These defects were all remedied by dividing the iron disc itself instead of the pitch; this could be done to any degree of minuteness required, and the continuous pitchy surface so reduced that its thickness might be made a minimum, in fact, not greater than necessary to satisfy the condition of enduring the small amount of abrasion which takes place during the time required to complete the polishing. The improvement which immediately followed this simple device was far greater than could have been anticipated, and the divided three-feet speculum, after this change, defined better with a power of 1200 than it had done before with a power of 300. Several polishers were made on this construction, the arrangement and dimensions of the grooves being somewhat different: that last used in polishing the two three-feet specula I think is the best, and a drawing of it has been annexed, figs. 8 and 9: the circular grooves were turned with a slide rest and are three-eighths of an inch deep and one quarter wide, leaving bands of continuous surface one quarter of an inch wide; the grooves at right angles are about one inch and a quarter apart, one quarter of an inch wide, and half an inch deep; they were cut with a small circular saw, under which the polisher was made to traverse on the bed of a large lathe.

The speculum was of course truly ground with the polisher first, and then the layer of pitch or resinous composition applied, the grooves remaining empty.

There are two conditions which I have found essential in producing a successful result; the one, that the polisher should fit the speculum exactly during the whole process; the other, that the resinous surface in contact with the speculum should be as hard as possible consistent with its admitting the polishing powder to imbed itself in it: without an attention to both of these, however accurately the motions of the machine may have been adjusted, however nearly a general parabolic figure may have been attained, the speculum will not define well. The first condition is satisfied by grooving the polisher, provided the resinous surface is sufficiently soft to expand laterally into the grooves when necessary; but when it is so, I have not found it hard enough to give a very true surface, and therefore the second condition is not satisfied. But here it is necessary to explain the meaning which I attach to the words *true surface*, in contradistinction to *accurate general figure*. A true surface is one which observes the law of continuity, when, in fact, the normal to the surface everywhere cuts the axis in conformity to the law of the curve, whatever that may be. In practice, the defect, which I call an untrue surface, is perceptible at a glance where it is very considerable, and the speculum is of long focus, for instance, twenty-seven feet; it is then only necessary to place the eye a few inches within the focus, while the speculum is turned to some bright object, as for instance the enamelled dial-plate of a watch, or the moon. The irregularity of the whole surface will then be apparent, more at the joints where the speculum is in separate pieces, but still more where there is a flaw or crack. The cause is obviously this: the metal under equal friction wears everywhere unequally, and therefore the inclination of the minute portions, I might almost say elements, of the surface, deviates slightly, but sensibly, and quite irregularly from the general curvature, producing an aberration independent either of general figure or aperture. A surface of speculum metal yields in the same irregular way to the action of acids, as indeed all metals do, but the more so as their texture is crystalline or fibrous. In proportion as the resinous surface is soft, and the polisher heavy, the irregularity increases, and therefore we should conclude that the harder the surface and lighter the polisher, the less the defect; and such is the fact. The accuracy of the *general figure* depends mainly upon the motions of the machine and the thinness of the resinous surface. If the resinous surface is so hard that the particles of polishing powder no longer sink into it deep enough to be held fast, then the polish is destroyed, the polishing process passing into that of grinding; long, however, before that limit of hardness has been attained, the resinous surface has lost its essential quality of expanding laterally, and therefore of preserving its exact coincidence of figure with the speculum. I have found that the two properties apparently inconsistent with each other, can be imparted to the polisher at the same time, simply by using the resinous composition of two different degrees of hardness, so as to form two very thin strata, the outer one being the harder. The resinous surface in contact with the speculum can thus be made as hard as necessary, while the thin subjacent layer of softer resin expands laterally, so as to preserve the figure of the polisher. The process of polishing for optical purposes appears to me in some measure to resemble that of filing, the polishing powder imbedded in the resinous surface representing the teeth of the file: while the polisher preserves its figure exactly, and consequently its contact with the speculum is exceedingly close, every particle of polishing powder, as it insinuates itself between the rubbing surfaces, must be instantly forced into the resin, deeper as the resin is softer, producing a grooving or grain in the speculum, finer, if the fineness of the polishing powder is given, in proportion to the softness of the resin, and, consequently, to the depth the particles have become imbedded, and, therefore, the smallness of the portion of each which projects; but the moment the figure of the polisher ceases to be exact, then the polishing powder is no longer forced into the resin, but runs loose, producing a grain perhaps as coarse, or coarser than the size of the particles of the powder itself. Hence, therefore, in practice it is of no less importance to the production of a fine polish, than it is to the production of a fine figure, that the polisher should very exactly fit the speculum during the whole operation. I find invariably that the moment that exact coincidence ceases, the polish rapidly declines, and is soon completely spoiled. I have hitherto observed that the quality of the polish which yields the maximum of defining power is that which is technically called a black polish, provided a very fine grain is perceptible when the speculum is placed near a window. A speculum may be polished so that the surface appears black, and without grain, like a surface of quicksilver; but I have always found it necessary for that purpose, to employ a softer resinous composition than seems consistent with the production of a very true surface. Conceiving that such a polish, though I did not find it reflected more light, was likely to reflect more accurately, I tried a vast number of experiments with the view of obtaining it in conjunction with the truest surface, but hitherto without success: the subject, however, perhaps, deserves to be pursued further, and as it seems impracticable without injury to soften the resinous surface, the best chance seems to be to search for some polishing substance consisting of smaller particles than the fine peroxide of iron, the one I have always used, so as to produce a grain not exceeding the magnitude which theory has assigned as that of an undulation of light. In preparing the resinous surface for the polisher, I have for a long time employed a mixture of common resin and turpentine, instead of pitch, having previously experienced much inconvenience in polishing large surfaces from the gritty particles which the pitch I was in the habit of using very frequently contained. However, whether pitch or resin be made use of, it is absolutely necessary that the hardness should be adjusted to the proper standard with great care.

Mr. MUDGE and Mr. EDWARDS have given different directions on this subject; Mr. MUDGE recommending the pitch to be rather soft, and Mr. EDWARDS very hard; but in the common mode of conducting the operation, no precaution being used to prevent the temperature of the speculum from progressively increasing from friction, at length a point will necessarily be attained, sooner or later, according to the hardness of the pitch, when the pitch will yield so as accurately to fit the speculum, and the polish will then rapidly improve: but should the temperature rise further before the polish is complete, the pitch will have become too soft to work a very true surface, and the speculum, even though in every other respect perfect, will not define very sharply. It is very evident that, under such circumstances, a steady temperature could not be maintained, except where the heat evolved by friction exactly balanced that dissipated by radiation and conduction, and that the result even on the limited scale of Mr. MUDGE's and Mr. EDWARDS's operations must have been uncertain; but in working large specula, the uncertainty was so great, that it gave rise to difficulties which I found it impossible to combat, and therefore I resorted to the simple expedient of making the speculum revolve in water, kept at an uniform temperature, generally 55°: all change also in the figure of the speculum, from variation of temperature during the process, was thus at the same time prevented. The hardness of the resinous surface was therefore adjusted to suit the temperature, which is thus easily effected. Common resin is melted, and when nearly boiling, spirit of turpentine is added to it, perhaps about one-fifth of its weight; but resin varies so much in quality, that there is no guide except actual trial. When the mixture has been incorporated by stirring, a cold piece of iron is to be immersed in it, and then placed for some minutes in a vessel of water at a temperature of 55°; if then a moderate pressure of the nail makes a decided impression without splintering, it is of a proper hardness for the first layer on the polisher, and only requires to be strained through canvas. I know of no mode which in practice answers better than the very rude one of judging of the hardness by the effects of the pressure of the thumb nail: there are others more precise, but they all take too much time, and sufficient accuracy can be attained without them. For the second layer, it is mixed with one-fourth of wheat flour, which, by increasing its tenacity and diminishing its adhesiveness, prevents that accident complained of by practical men, viz. the separation of minute particles of pitch from the polisher, which afterwards run loose between the polisher and the speculum. It is to be boiled till the water of the flour has been expelled, and the mixture becomes clear, and the boiling further continued till some of the turpentine has been driven off, and the mixture has become so hard, that at a temperature of 55°, a very strong pressure of the nail makes but a slight impression: it is still too soft, and I then add to it an equal weight of resin; it will then be hard enough to produce a very true surface, and, at the same time, soft enough to suffer the particles of polishing powder to imbed themselves, and consequently to produce a very fine black polish. Whenever the resinous mixture is remelted, I suspend the vessel to the beam of a scale, counterpoise it, and take care to apply the heat so gradually as not to drive off any of the turpentine, which is immediately perceptible by the disturbance of the equilibrium. To apply the resin, the polisher is first heated to about 150°, and the soft mixture laid on with a large flat brush, to about the thickness of one-thirtieth, or one-twenty-fifth of an inch; it is then suffered to cool to about 100°, and the hard mixture applied in the same way and to about the same thickness. When the temperature has sunk to 80°, the polisher is placed on the speculum previously covered with peroxide of iron and water, of about the consistence of thin cream. It may be well to observe, that the speculum is not endangered by applying a polisher at 80°, while it is but 55°, because the thin layer of resin retards very much the transmission of heat; but in grinding, where there is no resin, were there such a disparity, the speculum would be broken. In grinding the three-feet speculum formed in separate pieces, the iron plate happened to have been washed with warm water, and though but little warmer than the speculum, the moment it was put on several of the plates cracked, but from the construction it was but little injured: had it been the other speculum in one piece, it would of course have been destroyed.

I prepare the peroxide of iron by precipitation with water of ammonia from a pure dilute solution of sulphate of iron; the precipitate is washed, pressed in a screw press till nearly dry, and exposed to a heat which in the dark appears a dull low red. The only points of importance are, that the sulphate of iron should be pure, that the water of ammonia should be decidedly in excess, and that the heat should not exceed that I have described. The colour will be a bright crimson, inclining to yellow. I have tried both potass and soda pure instead of water of ammonia, but after washing with some degree of care, a trace of the alkali still remained, and the peroxide was of an ochrey colour till overheated, and did not polish properly. THOMSON says the peroxide of iron is sometimes of an ochrey colour, probably owing to some impurity, and I have found that the slightest trace of potass or soda produces that effect. Possibly even washing with a degree of care, too troublesome for practice, would be ineffectual in removing the last remains of the alkali, as DAVY found that silica prepared with an alkali always retained a trace of it, even after the most careful washing; but this is not exactly a case in point.

Having thus endeavoured to point out as concisely as possible what I have found essential in producing a *very true surface* and a fine polish, without at all wishing it to be inferred that I consider these processes quite perfect,—for, on the contrary, I believe much still remains to be accomplished,—I will next describe the means by which I have endeavoured to obtain a very good general figure.

When I had but little experience in working specula, considering the subject more theoretically than practically, I thought that a spherical figure was the only one which could be wrought with sufficient accuracy for optical purposes, and therefore that an original spherical aberration was an unavoidable evil; but that were it possible by any counteracting means to neutralize, or even diminish it, we might have telescopes of greater aperture with a given focal length. I constructed, therefore, a Newtonian telescope of six inches aperture, and two feet focus*: the speculum was in two concentric

* Sir DAVID BREWSTER's Journal, July, 1828.

MDCCCXL. 3 x

portions of such dimensions, that, presuming the original figure to have been accurately spherical, each should by calculation have one-half of the whole aberration : by drawing, therefore, the central portion back the proper calculated distance by a delicate screw adjustment, and bringing the images into exact coincidence, the aberration should have been reduced to one-half. The experiment was so far successful, that an instrument of eighteen inches aperture was commenced, and the castings of the speculum in three adjustible portions were completed. In the mean time, however, with additional experience, I found it necessary to adopt new views, and was soon convinced that it was not impossible to work figures which, though not rigidly accurate, were, however, better for specula than the spherical. The undertaking was, therefore, abandoned, but the original instrument still remains, and is so far curious as showing that an adjustment of such delicacy can be practically accomplished. The experiments to which I have alluded, were made with the elliptic polisher of Mr. EDWARDS, a contrivance in my opinion possessing more merit than has usually been ascribed to it : I found that a speculum of four inches aperture, and eighteen inches focus, after having been polished by hand as truly spherical as I could make it, was invariably improved by working it on the elliptic polisher ; however, on applying the same principle to larger specula, the result was less successful ; and after a great many trials with a speculum of eighteen inches aperture, I found it would not answer. On measuring the focal length of the surface at different distances from the centre of the *face*, it was certain that the radius of curvature always increased much too rapidly towards the edge ; and when the principle upon which the elliptical polisher acts is considered, it is evident that such a result might have been anticipated, and that the defect, though scarcely perceptible in very small specula, would have been very important where the dimensions were considerable.

Having observed that when the extent of the motions of the polishing machine were in certain proportions to the diameter of the speculum, its focal length gradually and regularly increased, that fact suggested another mode of working an approximate parabolic figure. If we suppose a spherical surface, under the operation of grinding and polishing, gradually to change into one of longer radius, it is very evident that, during that change, at no one instant of time will it be actually spherical, and the abrasion of the metal will be more rapid at each point as it is more distant from the centre of the *face*. When, however, the focal length neither increases nor diminishes, the abrasion will become uniform over the whole surface, producing a spherical figure. According, however, as the focal length (the actual average amount of abrasion during a given time being given) increases more or less rapidly, the nature of the curve will vary, and we might conceive it possible, having it in our power completely to control the rate at which the focal length increases, so to proportion the rate of that increase as to produce a surface approximating to that of the paraboloid. Of course the chances against obtaining an exact paraboloid are infinitely great, as an infinite number of curves may pass between the parabola and its circle of curvature,

H

and it is vain to look for a guide in searching for the proper one in calculations founded on the principles of exact science, as the effect of friction in polishing is not conformable to any known law; still from a number of experiments it might be possible to deduce an empirical formula practically valuable: this I have endeavoured to accomplish.

The weight of the polisher was constant, being the least possible consistent with its working properly, viz. ten pounds for a speculum three feet diameter. The distance of the counterpoising lever would obviously influence the curve; that I have regarded as constant also, viz. twelve feet, as also, in all my most recent experiments, the length of the stroke of the first eccentric B, which was one-third of the diameter of the speculum; the only variable quantity was, therefore, the stroke of the second eccentric G. Under these circumstances, the most accurate determination at which I have as yet been enabled to arrive is, that when the stroke of the second eccentric G is such as to communicate a lateral motion to the polisher equal to about ·27 of the diameter of the speculum, the curve will be nearly parabolic. The curvature I measure as MUDGE did, by means of diaphragms; and when the surface is true, the separate portions of it, though the general figure may be indifferent, will define sharply, and their focus can be ascertained with great precision. If the surface is not true, the curvature cannot be ascertained with any degree of accuracy; the watch-dial, the test object, is but ninety feet from the speculum, and could not be placed higher without inconvenience; it is therefore necessary in measuring the curvature of the speculum to allow for its distance. This I do simply by calculating the spherical aberration, the radiant being ninety feet distant, at two points, one $\frac{1}{9}$th of the radius, and the other $\frac{1}{3}$rd of the radius from the centre of the *face*, deducting that from the aberration for parallel rays at these points, and endeavouring so to figure the speculum that it should be over corrected that much. Any further refinement would be but waste of time in the present state of these experiments; and although we cannot hope to obtain anything but an approximation, still the limits of error will be so small that even a large fraction of that quantity may not in practice be very important. The adjustment, however, can be made with such accuracy, that the three-feet metal at present in the telescope, with its whole aperture, is thrown perceptibly out of focus by a motion of the eye-piece, amounting to less than the thirtieth of an inch; and even with a single lens of one-eighth of an inch focus, giving a power of 2592, the dots on a watch-dial are still in some degree defined.

A watch-dial is, upon the whole, as good a test for very large specula as can be desired, as there is so much light that the magnifying power can always be increased till indistinctness is perceptible; and although eventually the performance of the instrument on the heavenly bodies, the development of new details, or the discovery of new objects, which other instruments have not reached, are the proofs that an accession of instrumental power has been obtained, still as a test to have recourse to during the progress of experiments, a watch-dial, which is always at hand, and so

LORD OXMANTOWN ON THE REFLECTING TELESCOPE. 101

near that atmospheric changes do not very materially affect the result, is a far better object.

By keeping the same principles in view, a very perfect plane speculum can be worked with facility, difficult as that operation has been found in practice. The polisher I make use of for a metal about three inches by two, is three inches diameter, divided exactly like the large polisher, but with proportionate minuteness; when the metal is polished, it is tested in the usual way by viewing an object alternately by direct and reflected vision, with a very good thirty-inch achromatic, the aperture of which has been previously contracted to an inch and three quarters. If the metal is concave, it is worked with shorter strokes for about half an hour and then tried; it will be found to have become less concave, possibly convex; in the latter case it is to be worked with longer strokes; thus with the utmost facility a metal can be worked alternately concave and convex, and, with a little practice, the limit between the two can be hit with such exactness, that even with the severe test of a thirty-inch achromatic, no deviation from the plane can be perceived, and the loss of light will be the only evidence that the rays have suffered reflexion before their incidence on the object-glass. Smaller flat metals I find it better to polish on the same polisher, several together, according to their size.

The telescope is mounted very similarly to Sir JOHN HERSCHEL's, but in consequence of its greater size and weight, I have counterpoised both the tube and the whole machine, which makes it easily manageable; so that upon the whole, though, with the experience I now have, I believe the mounting might be improved, it is sufficiently convenient. I use it as a Newtonian, as I find that, with its large aperture and short focus, the saving of light by the Herschellian construction is not at all an equivalent for the sacrifice of defining power, at least that is the result of my present experience; the indistinctness, however, from the obliquity of the speculum, does not appear to me to be so great as I should have expected, considering the size of the circle of least confusion; for this I cannot account.

To prevent flexure in specula of moderate dimensions, I find it is quite sufficient to support them in their box on three strong iron plates, each plate being one-third part of a circular area, the same size as the speculum, and a sector of it; the plates rest at their centres of gravity on points fixed at the bottom of the box of the speculum, and therefore no flexure of the box can affect the speculum. Although the same simple means would probably be effectual for specula of the largest size, in supporting specula of three feet diameter I have availed myself of the suggestion of a clever Dublin artist, Mr. GRUBB, and, at the expense of a little more complication, have substituted nine plates for the three, resting on points supported by levers, which rest on three original points; and if flexure is thus more effectually prevented, which I think it ought to be, the additional workmanship is of no importance. This lever apparatus, however, must be exceedingly substantial, quite disproportionately so, otherwise tremors would be introduced by it, attended with the worst consequences.

In selecting the materials for this communication from a vast mass of experiments, I have endeavoured to collect together such results as were most likely to be useful, and to work the whole into a shape as practical as possible, compressing it at the same time into the smallest compass: there are many minor matters of detail, which, therefore, I have unavoidably omitted; some of them possibly might be valuable to men of science, who may be disposed to take up the subject practically with a view of proceeding further than I have done. I have, however, the less regret on this account, as, although the instrument and the laboratory where it was constructed are in the centre of Ireland, the facilities of communication are such, that those who may desire further information, can easily obtain it on the spot, and form their own estimate of the performance of the instrument.

Since the commencement of last September, when the telescope was completed and in perfect working order, till Christmas, all opportunities which presented themselves for observing, not very many indeed, were taken advantage of. A considerable number of Sir JOHN HERSCHEL's test objects were examined, and the performance of the instrument was quite satisfactory; but as to double stars, perhaps the most striking contrast between its action and that of other instruments, was the extreme brilliancy of the minute companions of large stars; for instance, the companion of Polaris, with six hundred, was very like Polaris itself in a forty-four inch achromatic, with a two and three quarter object glass. The companions of Alpha Lyræ, and Rigel, were brilliant objects. As to the nebulæ, though it was impossible not to feel persuaded that a larger, and equally perfect instrument would have done much more, still there was enough, I think, to justify a confident expectation, that even the present instrument will add something to the very little that is known respecting these wonderful bodies. I think I might almost venture to say that the nebulæ, 27 Messier, the annular nebula in Lyra, and what is perhaps more curious, the edge of the great nebula in Andromeda, have shown very evident symptoms of resolvability; as also other nebulæ less remarkable*. The appearance is that of a resolvable nebula in a telescope not quite powerful enough to resolve it completely. No such appearance, however, was observed till the power reached six hundred, and sometimes it was more decisive with powers of eight hundred and one thousand.

It is evident, therefore, that, except for the discovery of very faint nebulæ, an object, perhaps, of but little interest, nothing would be effected by constructing a telescope of the greatest dimensions, unless it was at the same time proportionately perfect; that a mere light grasper would do nothing.

This instrument acts very powerfully on the lunar surface, and, as might be supposed, shows everywhere a variety of details not marked in the beautiful map of BEER

---

* In describing the appearance of these bodies, I am anxious to guard myself from being supposed to consider it certain that they are actually resolvable, in the absence of that complete re-solution which leaves no room for error; nothing but the concurring opinion of several observers could in any degree impart to an inference the character of an astronomical fact.

and Mœdler. A much higher magnifying power can also be used with decided advantage than they say has hitherto been practicable, viz. 300*: with a power of 600, and sometimes with a power of 900, many details are brought out, not visible with lower powers.

Were it practicable to construct a rectangular prism, reflecting as accurately as a flat metal, notwithstanding the thickness of the glass, there should, I think, be a considerable saving of light: I have not as yet had time to try the experiment, but I think it worth a fair trial. Such prisms, however, do not seem to have performed as well as might have been expected, perhaps owing to imperfect workmanship.

A still greater accession of light might be obtained without sacrifice of defining power, by using the Herschellian construction, were it possible to discover some means of working approximately the surface of accurate reflexion for oblique rays. I have recently tried a few experiments on a small scale on this subject, and am disposed to think the task is not hopeless; but a course of experiments on a large scale would be required to afford a decisive result. The question is not whether such a figure could be worked as accurately as a spherical one; of that I have no hope; but whether, in practice, such a degree of accuracy might not be obtained, that the mirror would define decidedly better than if it had been spherical, and as well, or nearly so, as when worked to the best parabolic figure that can be executed, and used as a Newtonian. The principle on which I have proceeded in these experiments was simply to consider the reflecting surface sought, as a portion of a paraboloid whose axis coincided with the side of the tube, the eye-piece of course placed on the axis; and in endeavouring to work that figure, I have had recourse to no other expedient than an adjustment of the motions with respect to the position of the speculum, guided by the same view of the subject which directed the attempts to work the paraboloid for the Newtonian, varied merely to suit the altered circumstances; but in the present imperfect state of these experiments it would be waste of time to enter more into particulars. I have mentioned the subject merely for the purpose of directing the attention of others to it.

Sir William Hamilton, in his paper in the Transactions of the Royal Irish Academy for 1828, on Systems of Rays, considers the surface as the envelope of an ellipsoid of revolution, having a constant axis, but a variable eccentricity, moved in such a manner, that while one focus traverses in all directions the surface which cuts the incident rays perpendicularly, the other focus remains fixed at the point through which all the reflected rays are to pass. I have not, however, discovered any practical means of availing myself of his very original mode of treating the subject.

To conclude, I think I may state as the results of all these experiments, that specula can be made to act effectively, cast of the finest speculum metal, in separate portions,

* Bis jetzt ist eine 300 malige Vergrösserung die stärkste die man mit verhältnissmässigem Erfolge auf den Mond anwenden konnte, p. 5, *note.*

H—1

retained in their positions by an alloy of zinc and copper, as easily wrought as common brass, and that they can be executed in this manner of any required size; that castings of the finest speculum metal can be executed of large dimensions, perfect, and not very liable to break; that machinery can be employed with the greatest advantage in grinding and polishing specula; that, to obtain the finest polish, it is not necessary that the speculum should become warm, but that any temperature may be fixed upon and preserved uniform during the whole process; and that large specula can be polished as accurately as small ones, and be supported so as to be secured from flexure.

To form any other than a very vague estimate of the dimensions which the reflecting telescope may yet attain, would be impossible. Without allowing for further improvements in the process of polishing, which certainly may be confidently anticipated, I think that a speculum of six feet aperture could be made to bear a magnifying power more than sufficient to render the whole pencil of light available, and that in favourable states of the atmosphere it would act efficiently, without having recourse to the expedient which NEWTON pointed out as the last resource, that of observing from the summit of a high mountain. The construction, however, of such an instrument would be a serious task, and I should be sorry to attempt it, till, after additional experience in observing, and further opportunities of comparing the two three-feet specula already finished, I felt more competent to do justice to the undertaking. In the mean time, I hope to receive from scientific men suggestions, which would be most valuable; to continue the experiments already in progress, and to arrange the details of the mechanism necessary to render so large a tube conveniently manageable. Everything, then, having been previously determined with care, subsequent alterations would not be required; tedious experiments would not now be necessary, either in constructing the speculum, or in the less interesting but necessary task of acquiring a practical knowledge of the mechanic arts; and an instrument even of the gigantic dimensions I have proposed might, I think, be commenced and completed within one year.

Observations on some of the Nebulæ.   By the Earl of ROSSE, *F.R.S., &c.*

Received June 10,—Read June 13, 1844.

AS every addition, however trifling, to the little we know with certainty respecting the nebulæ can scarcely be considered wholly uninteresting, I have ventured to communicate a few observations made with the speculum of three feet aperture described in the Philosophical Transactions for 1840.

In using that instrument, the object was rather to test its powers, and to decide the merits of progressive experiments, than to seek for astronomical results, and therefore the micrometer was not employed; in every case, however, the sketches were repeatedly compared with the originals, and having usually the advantage of the opinion of one or more friends, and always of that of my assistant, I believe they will be found to be tolerably correct. Without accurate micrometrical measurements any sketch can be of comparatively little value as an astronomical record, since it would be scarcely safe under any circumstances to consider it as decisive evidence, where the question was whether any change had or had not taken place in the general outline or internal structure of a nebula, or in the relative positions of the remarkable stars in or near it. Still measurements to be valuable must be exact; and it would perhaps have been misspent time to have employed the micrometer with an instrument not well suited to the purpose, when there was a prospect of being under the necessity of repeating the same operations again, probably at a very short interval, in making a complete examination of the nebulæ with the instruments now in progress.

From these trifling sketches, however, we may perhaps faintly see some indications of the course which our speculations on the physical structure of the nebulæ are likely to take under the guidance of increasing information. Some estimate may also perhaps be made of the amount of knowledge to be gained by an examination of all the known nebulæ with instruments like the present telescope, which astronomers favourably circumstanced may construct without any very serious difficulty.

The actual time in one year during which a powerful telescope can be used effectively is so short, that where observations must be accompanied by sketches, the progress is necessarily slow; and it is still more so when the micrometrical measurement of faint objects becomes an essential portion of the work. Out of a considerable number of tolerably good working nights there are very few, and even then often but for a short time, when high magnifying powers can be employed; so that upon the whole a great deal cannot be accomplished by one instrument in a limited period. As to the present telescope, it has not been constantly employed. Unless the beginning of the night was favourable nothing was ever done; a regular system of observing would have been quite incompatible with the constant care and attention required in the progress of the more important work, the construction of a telescope of greater power, so that much fewer objects have been examined than would otherwise have been practicable; and even with the great facilities afforded by Sir John Herschel's invaluable catalogue, the whole amount of work done has been no more than the examination of about two-thirds of the figured nebulæ, and a few others in the general catalogue, many of them, or rather, perhaps, most of them, under circumstances but moderately favourable.

The sketches were originally made in the gallery of the telescope, and represent the objects placed as they appeared, not as they actually exist in space. I have copied them without alteration, not thinking that anything would be gained by placing them approximately in their true positions, while they would perhaps be less convenient for future re-examination and correction. The references are to Sir John Herschel's Catalogue.

Plate XVIII. fig. 88 is one of the many well-known clusters; I have selected it merely for the purpose of showing that in such objects we find no new feature, nothing which had not been seen with instruments of inferior power; the stars, of course, are more brilliant, more separated, and more numerous. I fear that no amount of optical power will make these objects better known to us, though perhaps exact measurements may bring out something.

Fig. 81 is also a cluster; we perceive in this, however, a considerable change of appearance; it is no longer an oval resolvable Nebula; we see resolvable filaments singularly disposed, springing principally from its southern extremity, and not, as is usual in clusters, irregularly in all directions. Probably greater power would bring out other filaments, and it would then assume the ordinary form of a cluster. It is studded with stars, mixed however with a nebulosity probably consisting of stars too minute to be recognized. It is an easy object, and I have shown it to many, and all have been at once struck with its remarkable aspect. Everything in the sketch can be seen under moderately favourable circumstances.

Plate XIX. fig. 26, on the contrary, is a difficult object; it requires an extremely fine night and a tolerably high power; it is then seen to consist of innumerable stars, mixed with nebulosity; and when we turn the eye from the telescope to the Milky Way, the similarity is so striking that it is impossible not to feel a pretty strong conviction that the nebulosity in both proceeds from the same cause.

Fig. 29.—The annular nebula in Lyra; 2 is the star in Sir John Herschel's sketch; I have inserted the six other stars as in some degree tests of the power of a telescope. Near star 3 there are two very minute stars seen with great difficulty; the others are easily seen whenever the night is sufficiently good to show the nebula well. The filaments proceeding from the edge become more conspicuous under increasing magnifying power within certain limits, which is strikingly characteristic of a cluster; still I do not feel confident that it is resolvable. I am however disposed to

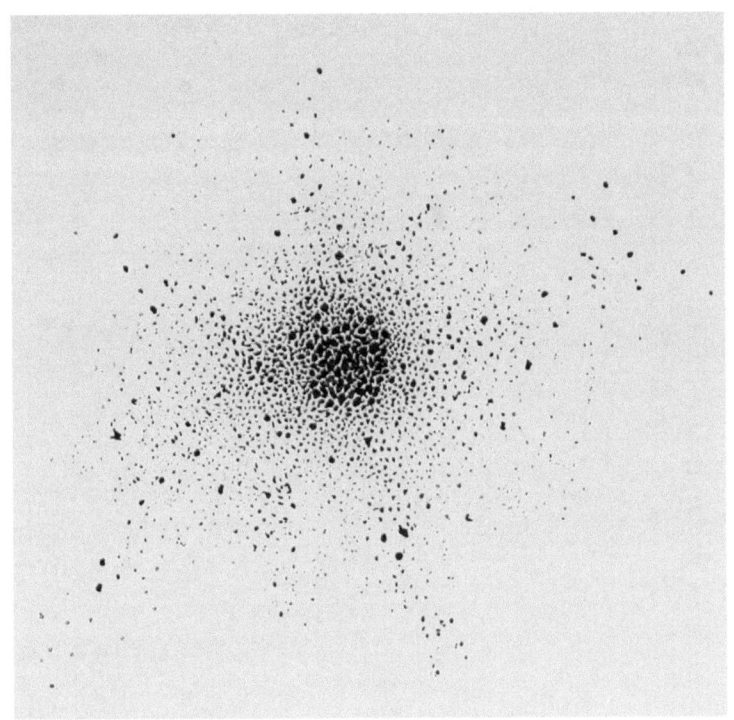

Fig. 88   R.A. 21ʰ 25ʹ
             Dec. 1° 34ʹ South

Fig. 81   R.A. 5ʰ 24
             Dec. 21° 53 North.

J Basire so

think that it was never examined when the instrument was in as good order, and the night as favourable, as on the several occasions when the resolvable character of fig. 26 was ascertained.

Fig. 47 is one apparently of another class. It has a star in the centre, and is of unequal brightness; the nebulosity is in patches, and I have sometimes fancied, though probably erroneously, that I could discover in it a faint resemblance to fig. 26. The star in the centre is easily seen, and there is nothing peculiar in its appearance; it is exactly like other stars seen in nebulæ; still it may really be but the brilliant condensed centre of a very remote cluster. I have not, however, detected any gradual increase of brilliancy towards the centre.

Not to multiply sketches which soon may require correction, I shall merely add that in fig. 32 we also find a star in the centre, and in fig. 85 likewise a star in the centre, and many other minute stars in and close to it, so that it is really a cluster. The double nebula (fig. 72) consists of two clusters, between which there is a star easily seen on even an indifferent night. In fig. 49 there are minute stars between and about the three large stars, and I think there can be no doubt it is a cluster. Fig. 25 abounds in stars mixed with nebulosity; I have not seen it on a very fine night, but it was observed by my assistant, and by a gentleman who was with him, and they had no doubt but that the centre was completely resolved. In the little annular nebula, fig. 48, I see nothing remarkable, farther than a star in the north preceding edge; it is tolerably conspicuous, and is about half-way between the exterior and interior circumference of the annulus.

Fig. 45 is a very remarkable object. It is no longer a planetary nebula, but an annular nebula, like that of Lyra, with a similarly fringed edge, though much less distinctly seen: it is oval, but the central portion is not so dark as that of Lyra; it very closely resembles the annular nebula of Lyra seen with an instrument of inferior power.

In several of the other figured nebulæ something has been discovered as to matters of detail. In some we have found perhaps a few minute and apparently accidental stars, in others a larger extent of nebulosity, and consequently a different form of outline, but nothing of sufficient importance to make it desirable further to prolong this paper. It appears to me, however, to be an important fact, that all we have seen strongly confirms the accuracy of Sir John Herschel's judgement in selecting the nebulæ which he places in the class designated as resolvable. It is important from its bearing on future researches; for where the power of our instruments is insufficient to do more than to bring to light distinctly the peculiar characteristics of resolvability, these once observed with due caution and their reality ascertained beyond doubt, we shall conclude with little danger of error, that the object is really a cluster. We should err, however, were we to assume the converse of the proposition, that the absence of all symptoms of resolvability was evidence conclusive that the object was not a cluster. In some instances, with increasing optical power, the resolvable character has become clearly developed, as in fig. 26, and a further increase of power has shown the object resolved.

It is also perhaps important to observe, that now, as has always been the case, an increase of instrumental power has added to the number of the clusters at the expense of the nebulæ, properly so called; still it would be very unsafe to conclude that such will always be the case, and thence to draw the obvious inference that all nebulosity is but the glare of stars too remote to be separated by the utmost power of our instruments.

The magnifying powers I have usually employed vary from 250 to 800; occasionally much higher powers have been useful; but to see every thing described in this paper, a power of 600 with perfect definition is sufficient.

In my paper on the Reflecting Telescope, in the Philosophical Transactions for 1840, there is an error which I am anxious to take this opportunity of correcting. In page 523, line 14, instead of "to the polisher," it should have been "to the bar which moves the polisher measured on the edge F of the tank." We have always estimated the effect of the excentric G by the space on the edge of the tank traversed by the bar: the number 2·7 was entered down at once from the journal-book, whereas, to have suited it to the form of expression made use of, it should have been reduced to the centre of the polisher.

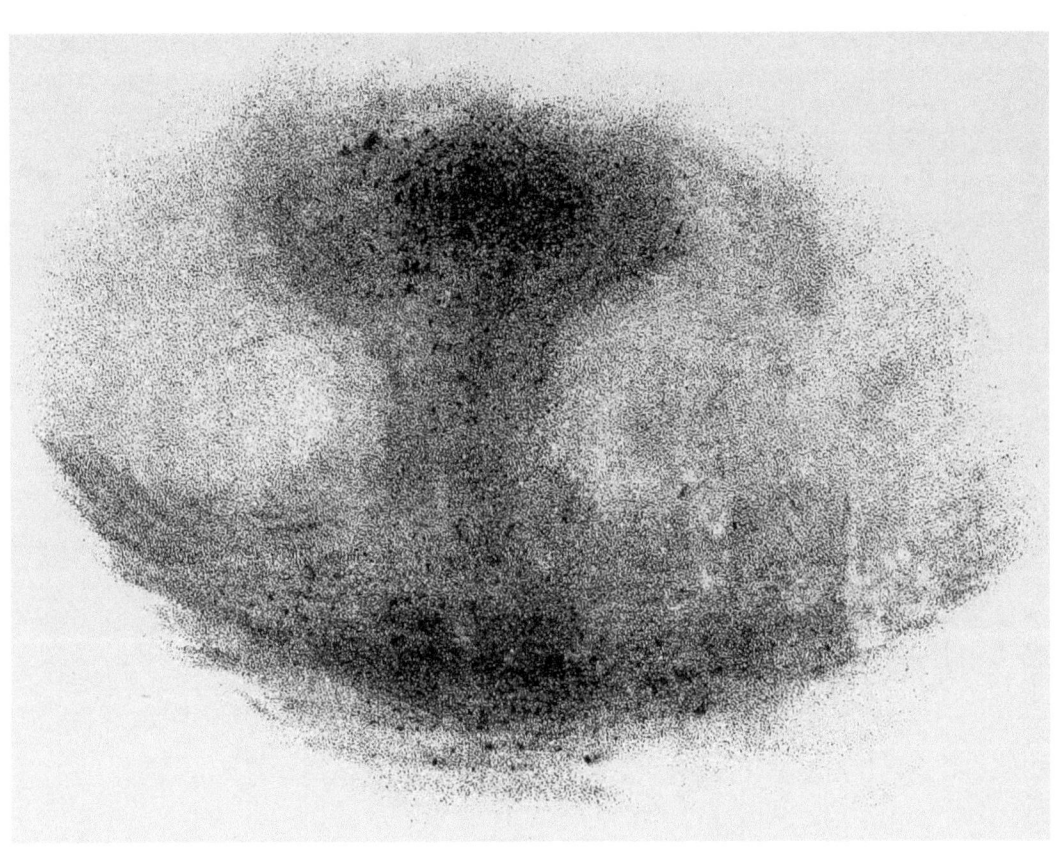

Fig. 26.   R.A. 19ʰ 52´
           Dec. 32° 49´ North.

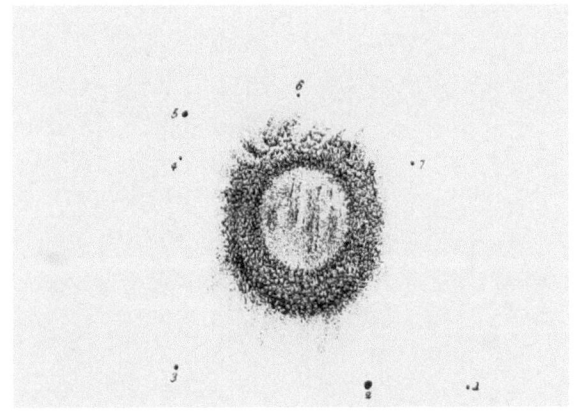

Fig. 29   R.A. 18ʰ 47´
          Dec. 32° 49´ North.

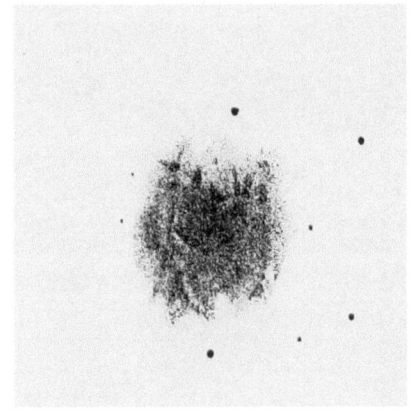

Fig. 47   R.A. 20ʰ 15´
          Dec. 19° 34´ North.

XXV. *Observations on the Nebulæ. By The Earl of* Rosse, *Pres. R.S., &c. &c.*

Received June 19,—Read June 20, 1850.

IN laying before the Royal Society an account of the progress which has been made up to the present date in the re-examination of Sir John Herschel's Catalogue of Nebulæ published in the Philosophical Transactions for 1833, it will be necessary to say something of the qualities of the instrument employed.

The telescope has a clear aperture of 6 feet, and a focal length of 53 feet. It has hitherto been used as a Newtonian, but in constructing the galleries provision was made for the easy application of a little additional apparatus to change the height of the observer, so that the focal length of the speculum remaining the same, the instrument could be conveniently worked as a Herschelian.

Although with an aperture so great in proportion to the focal length, the performance of a parabolic speculum placed obliquely would no doubt be very unsatisfactory, still additional light is so important in bringing out faint details, that it is not improbable in the further examination of the objects of most promise with the full light of the speculum, *undiminished by a second reflexion*, some additional features of interest will come out.

The second reflexion is accomplished in the usual way by a surface of speculum metal; some experiments have been made, suggested by Jamin's paper in the Annales de Chimie for 1848, to procure a surface of silver suited to the purpose, but without complete success. Arrangements also have been for some time in contemplation with the view of effecting the second reflexion occasionally by a small glass prism; and about a year ago a prism was procured from Munich for the purpose: in both cases there would be a great saving of light; but I am speaking of the instrument as it is, not as it may become, if further improved.

The tube reposes at its lower end upon a very massive universal joint of cast iron, resting on a pier of stonework buried in the ground; and it is counterpoised so that it can be moved in polar distance with great facility. A quick motion in polar distance is given by a windlass below, and a slow motion is given by hand above for measurements. The extreme range of the tube in right ascension at the equator is one hour; but greater as the polar distance diminishes. The quick movement in right ascension is given below by a wheel turned by a workman, and the slow motion by hand above; the instrument is therefore completely under the dominion of the observer. The tube is slung entirely by chains, and is perfectly steady even in a gale of wind.

As the chain which governs the movement of the telescope passes over a pulley capable of being brought by a little subsidiary apparatus into a line drawn from the axis of motion parallel to the axis of the earth, the movement of the telescope can be rendered almost exactly equatorial : there was some mechanical advantage in placing the pulley a little out of that line ; and for such measurements as we have required, we have found the movement of the telescope sufficiently equatorial without the subsidiary apparatus, and therefore have not up to the present time made use of it. When the telescope is in the meridian, as it moves in polar distance it is guided by a cast-iron arc of a circle about 85 feet diameter nicely planed. The arc is composed of pieces 5 feet long, each adjusted independently in the meridian by the transit instrument, and secured to massive stonework. The horizontal axis of the great universal joint gives motion to an index which points to polar distances on an arc of 6 feet radius, by which the telescope is very quickly set in polar distance. A 20-inch circle with a very delicate level, attached to the telescope, performs the same office, more slowly but with greater accuracy ; and also gives polar distances with considerable precision when duly corrected. The whole mounting was planned especially with a view of carrying on a regular system of sweeping, for which it is peculiarly adapted ; but the known objects which require examination are so numerous that hitherto we have been fully occupied with them ; and the discovery of new nebulæ has as yet formed no part of the systematic work of the observatory.

As yet the telescope is not provided with a clock movement. A clock movement was part of the original design, and there would have been no serious difficulty in carrying it out ; but the want of it has not been very much felt, and there were other matters requiring more immediate attention.

Various micrometers have been tried, but upon the whole the common wire micrometer with thick lines succeeds the best. The thick lines are formed by coiling very fine silver wire four times round the forks, soldering it there, and then removing the lower half of the coil. A little spirit varnish unites the fine wires into a thin ribbon with a straight edge, perhaps as perfect as can be made. The micrometer is used without illumination ; and I have never failed to see the lines in the darkest night ; but of course measurements with thick lines are inferior in point of accuracy to measurements with thin lines in an illuminated field. Unfortunately any micrometrical contrivance which either diminishes the light of the telescope, or renders the field less dark, extinguishes the faint details of the nebulæ, which even with an aperture of 6 feet are often barely perceptible. There have been many ingenious attempts to make fine lines visible in a perfectly dark field, but they have not, at least as far as my experience goes, been entirely successful.

The telescope has two specula, one about three and a half, and the other a little more than four tons weight. Each speculum was originally provided with a system of levers to afford it an equable support : it was placed upon this system before it was ground, and it has rested upon it ever since. The system of levers is a combination of three systems in every respect similar, resting on three points under the centres of gravity of the three equal sectors into which the speculum may be supposed to be

Fig. 1.

Phil. Trans. MDCCCL. *Plate* XXXV. *p.* 199.

Fig. 2.

J. Basire sc.

divided. Each system consists of one triangle with its point of support directly under its centre of gravity, upon which it freely oscillates. This triangle carries at its angles three similar points of support for three other triangles, under their centres of gravity, and they again at their angles carry in a similar way cast-iron platforms formed of thin ribs so as to make a kind of irregular open-work grating, supported under their centres of gravity. These platforms are all of equal area though not of similar shape. As there are three systems there are therefore twenty-seven platforms, which together make a circular disc about an inch in diameter less than the speculum: when arranged however a little apart so as not to touch, they make a disc about the same diameter as the speculum. Each platform is coated with greased cloth, and may be considered as bearing up one twenty-seventh of the weight of the speculum. Between the platforms and the speculum pieces of tin plate are inserted to diminish the friction as much as possible. The platforms being of open-work, they do not prevent the water in which the speculum is immersed from freely carrying away the heat as it is developed during the process of polishing, which is essential.

It is evident that a speculum so supported will be practically free from strain while in a horizontal position, provided the due action of the levers is not interfered with by any disturbing force; it will be very much in the same condition as if floating in a vessel of mercury; when it ceases to be horizontal however new forces come into play: part of the weight must then be resisted by pressure against the edge. Four very strong segments of cast iron, each about one-eighth of the circumference, were adjusted to the edge by screws, the segments bearing upon the massive castings which sustained the three primary supports of the lever apparatus. Provision was made to allow a little motion perpendicular to the plane of the speculum, to guard as much as possible against strain from the elasticity of the lever apparatus, which was however very small, the yielding being less than one-fortieth of an inch.

The two specula of 3 feet aperture I have so long employed are mounted on a similar principle: they have however fewer points of support, and by a little sacrifice of the condition of perfect equilibrium, the whole system of levers was thrown without difficulty almost exactly into one plane. They are free from perceptible flexure in the different positions of the instrument. With the two specula of 6 feet diameter the case was otherwise. The 3-feet specula, weighing each about thirteen hundred weight, were very much stiffer, in proportion to their weight, than the 6-feet specula. To have made the 6-feet specula of equal proportionate stiffness, either they should have been enormously heavy, or the material should have been so disposed as to give greater stiffness than when simply cast into the shape of a solid disc. Some years ago it was ascertained by experiments, but on a small scale, that it would be practicable to dispose of three-fourths of the material of a speculum so as to secure a great increase of stiffness; the form adopted was a system of hexagonal cells. Whether on a great scale the difficulties would be too serious to be surmounted is a question; however it is with solid discs we have had to deal. The relative stiffness of speculum metal and wrought iron is about five to six three-tenths; yet strange as it may appear, so delicate is the optical test, that strong pressure of the hand at the back of a speculum, four tons weight, and nearly six inches thick, produces flexure sufficient to distort the image of a star. It is obvious, therefore, that a slight inequality in the action of the lever apparatus supporting a 6-feet speculum would produce an amount of flexure sufficient to destroy definition. It has not been found possible so to secure the 6-feet specula as to prevent a slight change of place in a plane parallel to the plane of the levers, and as the levers are not all in one plane as in the case of the 3-feet specula, and a considerable amount of friction exists between the speculum and its lever supports, when the speculum changes place, however slightly, there will be a force tending to disturb the equal actions of the levers. It has been found that when the speculum changes its place one-thirtieth of an inch, still adhering to its levers, unmistakeable distortion will be produced. We have occasionally observed, even during a night's work, the sudden appearance, and the as sudden disappearance of the rudiments of focal lines, the undoubted evidences of flexure; but we have not found that flexure, even to the extent of materially disfiguring the image of a large star, interferes much with the action of the speculum on the faint details of nebulæ, although it greatly lessens its power in bringing out minute points of light, and in showing resolvability where under favourable circumstances resolution had been previously effected.

In the spring of 1848 the heavier of the two specula for nearly three months performed admirably, very rarely exhibiting the slightest indication of flexure. It then remained inactive for some time before and after the solstice, and when we again commenced observing it was found to be in a state of strain; the friction between the lever apparatus and the speculum had no doubt in the meantime increased considerably, and the levers being therefore unable to adjust themselves to some slight but permanent change in the place of the speculum, they no longer supported it equably. It was cautiously raised a little by screws for the purpose of re-adjusting the levers, and to our surprise the unequal strain of the screws was found to have produced permanent flexure, so that the speculum did not again perform well till after it had been reground. From the experiments of Mr. EATON HODGKINSON and others, we should have been prepared for a change of figure in a mass of cast iron, but with a material so brittle and so elastic as speculum metal, the result was quite unexpected. Recently, in supporting the lighter of the two specula, twenty-seven triangles have been substituted for the twenty-seven platforms, each triangle carrying at its angles three brass balls, so that the speculum rolls freely on eighty-one balls, which support it pretty nearly equably. This appears to be a great improvement, but I will not dwell further on the subject. To describe the experiments which have been made with a view of discovering the best means of supporting very large specula, a question of great theoretical and practical difficulty, would occupy too much space, and would require elaborate engravings; it would besides be foreign to the object of this paper.

Fig 3

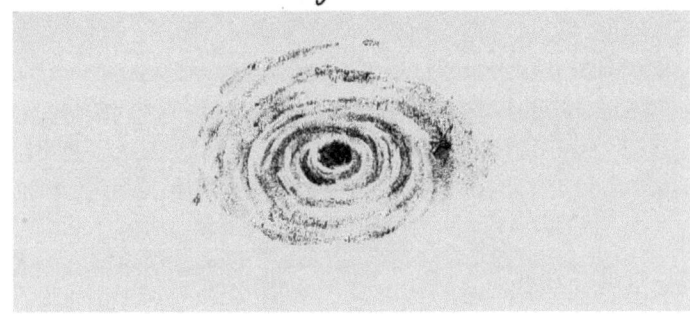

Fig 4

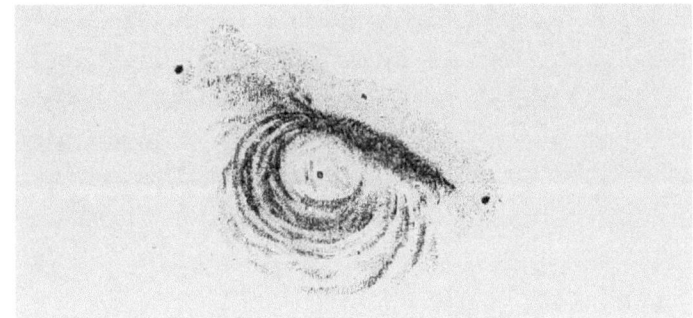

Fig. 5

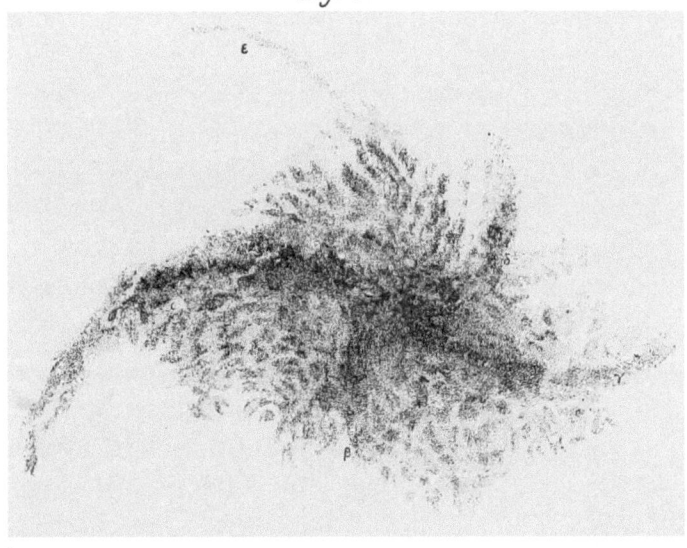

J Basire sc

The same considerations also forbid any more minute description of the telescope and its mounting.

From what has been said, it is evident that the 6-feet specula being occasionally in a state of strain, were not uniform in their action. There was however another cause of unequal action. The 6-feet specula, after they have been polished, cannot be tested till they have been removed from the laboratory to the telescope, there to await a good night, the great focal length making it impossible to test them while on the engine. Now it has often happened that a speculum which has subsequently proved to be incapable of very fine definition, has remained in the telescope during a succession of moderately good nights, when a great deal of work was done, awaiting a night when the air was in a state to warrant a decisive opinion. Such a speculum might do good work, but it would not resolve difficult nebulæ, neither would it bring out faint points of light, even when wide apart. There is still another cause of the unequal action of our specula far more serious, the varying state of the atmosphere. When the air is unsteady, minute stars are no longer points, the diffused image is much fainter, and single stars, easily seen when the air is steady, are no longer visible. When many minute stars are crowded together the whole become blended, and instead of a resolved nebula we have merely a diffused, perhaps bright nebulosity. The transparency of the air varies also quite as much; and the aspect of the nebulæ changes from night to night, just as the appearance of a distant building alters as the details of the architecture are more or less obscured by the intervening mist. With these facts, the Society will not be surprised should it be in our power at a future time to communicate some additional particulars, even as to the nebulæ which have been the most frequently observed.

The sketches which accompany this paper are on a very small scale, but they are sufficient to convey a pretty accurate idea of the peculiarities of structure which have gradually become known to us: in many of the nebulæ they are very remarkable, and seem even to indicate the presence of dynamical laws we may perhaps fancy to be almost within our grasp. To have made full-sized copies of the original sketches would have been useless, as many micrometrical measures are still wanting, and there are many matters of detail to be worked in before they will be entitled to rank as astronomical records, to be referred to as evidence of change, should there hereafter be any reason to suspect it.

Much however as the discovery of these strange forms may be calculated to excite our curiosity, and to awaken an intense desire to learn something of the laws which give order to these wonderful systems, as yet, I think, we have no fair ground even for plausible conjecture; and as observations have accumulated the subject has become, to my mind at least, more mysterious and more inapproachable. There has therefore been little temptation to indulge in speculation, and consequently there can have been but little danger of bias in seeking for the facts. When certain phenomena can only be seen with great difficulty, the eye may imperceptibly be in some degree influenced by the mind; therefore a preconceived theory may mislead, and speculations are not without danger. On the other hand, speculations may render important service by directing attention to phenomena which otherwise would escape observation, just as we are sometimes enabled to recognize a faint object with a small instrument, having had our attention previously directed to it by an instrument of greater power. The conjectures therefore of men of science are always to be invited as aids during the active prosecution of research.

It will be at once remarked, that the spiral arrangement so strongly developed in Plate XXXV. H. 1622, 51 MESSIER, fig. 1, is traceable, more or less distinctly, in several of the sketches. More frequently indeed there is a nearer approach to a kind of irregular interrupted annular disposition of the luminous material than to the regularity so striking in 51 MESSIER; but it can scarcely be doubted that these nebulæ are systems of a very similar nature, seen more or less perfectly, and variously placed to the line of sight. In general the details which characterize objects of this class are extremely faint, scarcely perhaps to be seen with certainty on a moderately good night with less than the full aperture of 6 feet: in 51 MESSIER, however, and perhaps a few more, it is not so. A 6-feet aperture so strikingly brings out the characteristic features of 51 MESSIER, that I think considerably less power would suffice, on a very fine night, to bring out the principal convolutions. This nebula has been seen by a great many visitors, and its general resemblance to the sketch at once recognized even by unpractised eyes. MESSIER describes this object as a double nebula without stars; Sir WILLIAM HERSCHEL as a bright round nebula, surrounded by a halo or glory at a distance from it, and accompanied by a companion; and Sir JOHN HERSCHEL observed the partial subdivision of the *s. f.* limb of the ring into two branches. Taking Sir J. HERSCHEL's figure, and placing it as it would be if seen with a Newtonian telescope, we shall at once recognise the bright convolutions of the spiral, which were seen by him as a divided ring. We thus observe, that with each successive increase of optical power, the structure has become more complicated and more unlike anything which we could picture to ourselves as the result of any form of dynamical law, of which we find a counterpart in our system. The connection of the companion with the greater nebula, of which there is not the least doubt, and in the way represented in the sketch, adds, as it appears to me, if possible, to the difficulty of forming any conceivable hypothesis. That such a system should exist, without internal movement, seems to be in the highest degree improbable: we may possibly aid our conceptions by coupling with the idea of motion that of a resisting medium; but we cannot regard such a system in any way as a case of mere statical equilibrium. Measurements therefore are of the highest interest, but unfortunately they are attended with great difficulties. Measurements of the points of maximum brightness in the motling of the different convolutions must necessarily be very loose; for although on the finest nights we see them breaking up into stars, the exceedingly minute stars cannot be seen steadily, and to identify one in each case would be im-

Fig. 9.

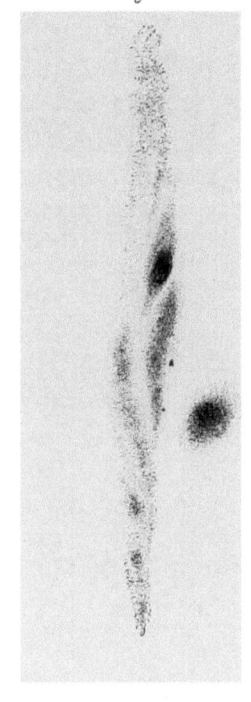

Fig 6.

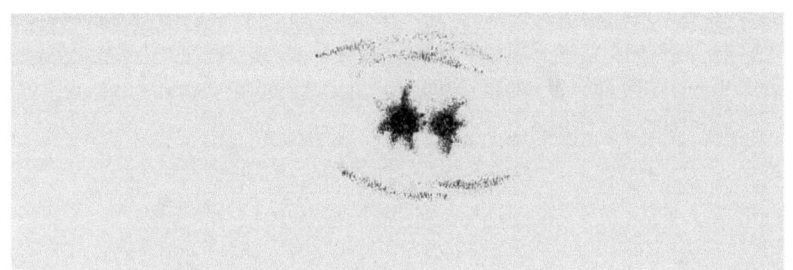

Fig 7

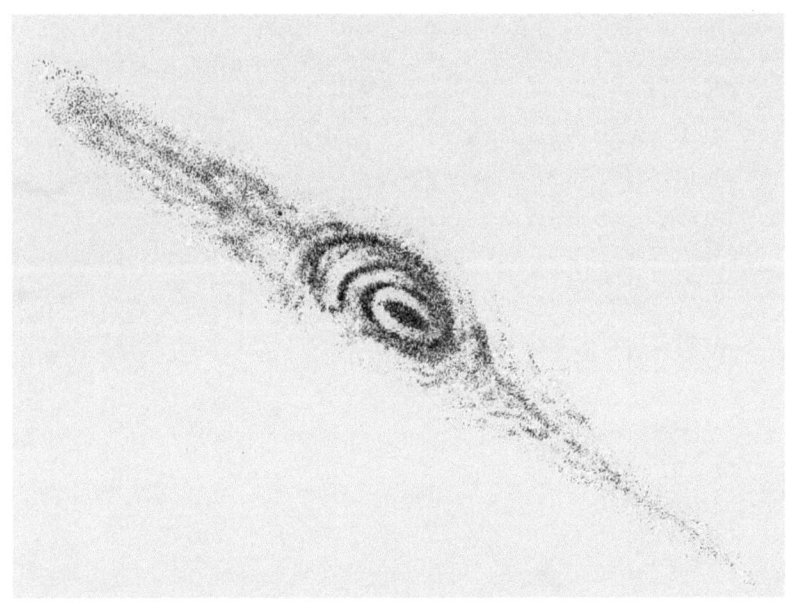

Fig. 10.

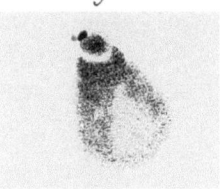

Fig 8.

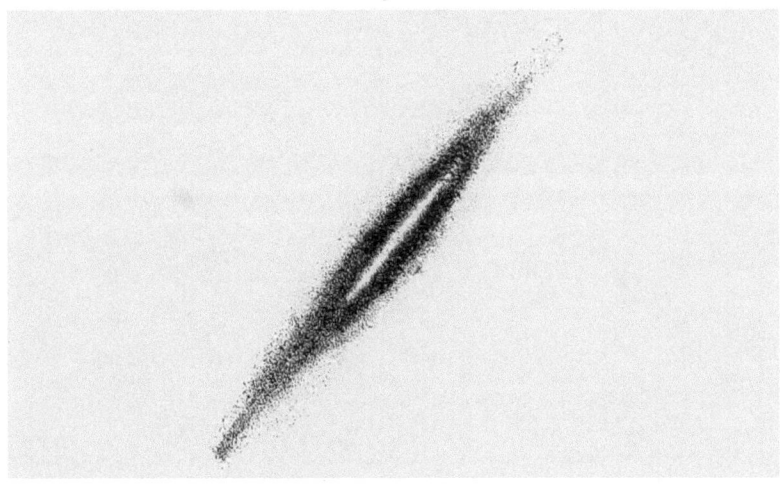

Fig. 11.

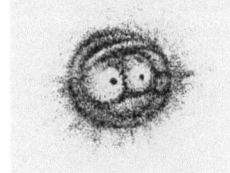

J. Basire sc

possible with our present means.   The nebula itself, however, is pretty well studded with stars, which can be distinctly seen of various sizes, and of a few of these, with reference to the principal nucleus, measurements were taken by my assistant, Mr. JOHNSTONE STONEY, in the spring of 1849, during my absence in London; for some time before the weather had been continually cloudy.   These measurements have been again repeated by him this year, 1850, during the months of April and May.   Just as was the case last year, in February and March the sky was almost constantly overcast.   He has also taken some measures from the centre of the principal nucleus to the apparent boundary of the coils, in different angles of position.   The micrometer employed was furnished with broad lines formed of a coil of silver wire in the way I have described, seen without illumination.   Some of the stars in the nebula are so bright, I have little doubt they would bear illumination; if so, their positions with respect to some one star might be obtained with great accuracy of course by employing spiders' lines; this season however it is too late to make the attempt.   Several of these stars are no doubt within the reach of the great instruments at Pulkova and at Cambridge, U.S., and I hope the distinguished astronomers who have charge of them will consider the subject worthy of their attention.   Their better climate gives them many advantages, of which not the least is the opportunity of devoting time to measurements without any serious interruption to other work.   I need perhaps hardly add, that measurements taken from the estimated centre of a nucleus, and still more from the estimated termination of nebulosity, are but the roughest approximations; they are however the only measurements nebulosity admits of, and if sufficiently numerous, I think they will bring to light any considerable change of place, or form, which may occur.

The spiral arrangement of 51 MESSIER was detected in the spring of 1845.   In the following spring an arrangement, also spiral but of a different character, was detected in 99 MESSIER, Plate XXXV fig. 2.   This object is also easily seen, and probably a smaller instrument, under favourable circumstances, would show everything in the sketch.   Numbers 3239 and 2370 of HERSCHEL's Southern Catalogue are very probably objects of a similar character, and as the same instrument does not seem to have revealed any trace of the form of 99 MESSIER, they are no doubt much more conspicuous.   It is not therefore unreasonable to hope, that whenever the southern hemisphere shall be re-examined with instruments of great power, these two remarkable nebulæ will yield some interesting result.

The other spiral nebulæ discovered up to the present time are comparatively difficult to be seen, and the full power of the instrument is required, at least in our climate, to bring out the details.   It should be observed that we are in the habit of calling all objects spirals in which we have detected a curvilinear arrangement not consisting of regular re-entering curves; it is convenient to class them under a common name, though we have not the means of proving that they are similar systems. They at present amount to fourteen, four of which have been discovered this spring: there are besides other nebulæ in which indications of the same character have been observed, but they are still marked doubtful in our working list, having been seen when the air was not very transparent; 51 MESSIER, Plate XXXV. fig. 1, is the most conspicuous object of that class.

The question may perhaps suggest itself whether there is not something in the aspect of a spiral nebula, which forces upon us the conviction that it is a system with an organization quite different from that of any known cluster.   The only answer I am enabled to give to that question is, that in the exterior stars of some clusters there appears to be a tendency to an arrangement in curved branches, which cannot well be unreal, or accidental.   Nos. 480, 1916, 1968, 1972, are the objects in which I observe that peculiarity noted down in our list of observations as suspected.   As to 1968, Sir JOHN HERSCHEL uses the following words in his Catalogue, " has hairy-looking curvilinear branches."   Careful drawings based on measurements would settle the question, whether the suspected curvilinear distribution of the stars is real or not; this would also perhaps settle another question of interest, whether the distribution of the stars in these objects is reconcileable with the hypothesis of an equal distribution of the stars of the system ; as yet however there has not been time to make the required measurements.   In passing from the spiral to the regular annular nebulæ, we perceive we are at once engaged with objects of a very different character : still here even there seems to be something like a connecting link ; the great round planetary nebula, H 838, Plate XXXVII. fig. 11, with a double perforation appears to partake of the structure both of the annular and spiral nebulæ.   There were but two annular nebulæ known in the northern hemisphere when Sir JOHN HERSCHEL's Catalogue was published ; now there are seven, as we have found that five of the planetary nebulæ are really annular. Of these objects, the annular nebula in Lyra is the one in which the form is by far the most easily recognized.   I have not yet sketched it with the 6-feet instrument, because I have never seen it under favourable circumstances : the opportunities of observing it well on the meridian are comparatively rare owing to twilight.   It was however observed seven times in 1848 and once in 1849.   The only additional particulars I collect from the observations, are that the central opening has considerably more nebulosity in it than it appeared to have with the 3-feet instrument, and that there is one pretty bright star in it, s.f. the centre, and a few other very minute stars.   In the sky round the nebula and near it there are several very small stars which were not before seen, and therefore the stars in the dark opening may possibly be merely accidental.   In the annulus, especially at the extremities of the minor axis, there are several minute stars, but there was still much nebulosity not seen as distinct stars.

The other annular nebula of HERSCHEL's Northern Catalogue is a much fainter object : it has been observed but once with the large instrument, August 1, 1848 ; but the evidence of resolution appears to have been more complete ; many stars were seen in the annulus ; one of them was very conspicuous.   That a faint nebula should be more easily resolvable than a bright one is not unusual, neither is it contrary to probability ; faintness may be owing to distance, or to a wider separation of the stars,

Fig. 12.

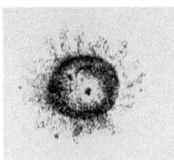

Fig. 15.

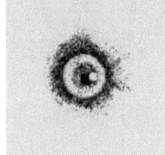

Fig. 13.

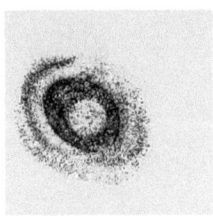

Fig. 16.

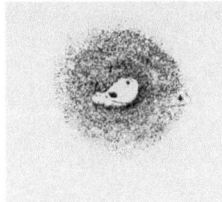

Fig. 14.

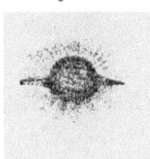

Fig. 17.

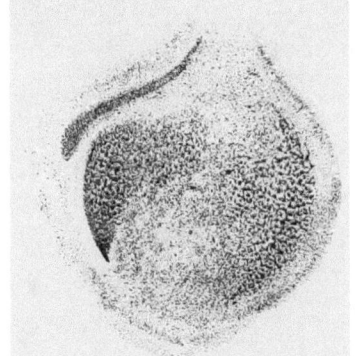

J. Baswo so

either physically or optically; in the latter case it is not unlikely that in a faint nebula they might be seen separate with an instrument of great aperture, while in the brighter and more closely packed nebula they were blended together, owing to imperfect definition, arising out of the state of the air, or instrument. As an example, the dumb-bell is a bright nebula: on three exceedingly fine nights succeeding each other at short intervals, the stars in the brighter parts of the nebula were better shown with 3 feet aperture than they have since been with 6 feet. Very fine nights, when the air seems to set no limits to magnifying power, are extremely rare, and the dumb-bell has not been seen with the great instrument on such nights. On the other hand, on all ordinary nights, a variety of details are shown by the great instrument which were not seen on the finest nights with the smaller instrument. There is another fact I may perhaps add, that while high magnifying power brings out minute stars it extinguishes faint nebulosity. The optical reason is obvious; but in sketching the dumb-bell nebula in 1845 that fact was overlooked, and but one eye-piece was used, a very high one; had there been a low one also used the sketch would have been more complete. To return to the annular nebulæ. The five planetary nebulæ we have ascertained to be annular, are as follows: 464, Plate XXXVIII. fig. 12, has two stars within it; 2075 has one star a little following the centre; 2241, Plate XXXVIII. fig. 13, has no star, but is surrounded with a faint external annulus; 2050 has a perforation not round nor quite symmetrical with the star; 838, Plate XXXVII. fig. 11, has two stars and two perforations. In no instance is the central opening quite dark. The planetary nebula, 2047, is marked in our journal as annular, but the observation is without date and other particulars, and therefore I do not consider it altogether trustworthy. In 2098, Plate XXXVIII. fig. 14, another planetary nebula, we have not detected any perforation, but it has ansæ, which probably indicate a surrounding nebulous ring seen edgeways, just as 450, Plate XXXVIII. fig. 15, has apparently a nebulous ring seen on the flat; and if the annular nebulæ are really hollow shells, the nebulous ring would cover the comparatively transparent centre; 365 and 2037 have never been observed.

Passing from the annular nebulæ to the nebulous stars, there are two objects well-worthy of especial notice.

Sir JOHN HERSCHEL very accurately describes a nebulous star thus:—"A sharp and brilliant star concentrically surrounded by a perfectly circular disc or atmosphere of faint light, in some cases dying away insensibly on all sides, in others almost suddenly terminated*." No. 450 of Sir JOHN HERSCHEL's Catalogue is one of these nebulous stars, and is there thus described:—"A star of the 8th magnitude, exactly in the centre of an exactly round and bright atmosphere, 25″ diameter. The star is quite stellar, not a mere nucleus. Another star, 8th magnitude, distant 100″, and about 85° n p, has no such atmosphere.—A most remarkable object."

Plate XXXVIII. fig. 15 represents this wonderful object as seen with the 6-feet telescope. It has been several times examined, and as yet we have not seen the

* Outlines of Astronomy, p. 605.

3 T 2

slightest indication of resolvability. The outer ring is seen on a pretty good night completely separated from the nucleus surrounding the brilliant point or star. The light is very bright, and always appeared to be flickering, owing no doubt to the unsteadiness of the atmosphere. There is a small dark space to the right of the star, which indicates a perforation similar perhaps to that discovered in Nos. 838, 2050, and others. The annular form of this object was detected by Mr. JOHNSTONE STONEY, my assistant, when observing alone, and the sketch is his; I have however since had ample opportunities of satisfying myself that the object has been accurately represented. Plate XXXVIII. fig. 16 represents the other nebulous star, Orionis: the remarkable feature in this object, the dark cavity, not symmetrical with the star, was also discovered by Mr. JOHNSTONE STONEY when observing alone with the 3-feet telescope: I have since seen it several times and sketched it. The components of Orionis have not been laid down micrometrically, or even with care by the eye, but the dark cavity with respect to the stars is faithfully represented. If the dark cavity was symmetrical with the stars, it might perhaps be thought by some that the phenomenon was optical, but as it is the thing is impossible. A small double star nf has similar openings, but they are not so easily seen. These openings appear to be of the same character as the opening within the bright stars of the trapezium of Orion, the stars being at the edges of the opening. Had the stars been situated all together within the openings, the suspicion would perhaps have suggested itself more strongly that the nebula had been absorbed by the stars. As it is, I think we can hardly fail to conclude that the nebula is in some way connected with these bright stars, in fact that they are equidistant, and therefore, if the inquiries about parallax, now proceeding with so much activity, should result in giving us the distances of these bright stars, we shall have the distance of this nebula.

The long elliptic or lenticular nebulæ are very numerous; I have given three sketches of remarkable objects of this class: the appearance of Plate XXXVII. fig. 7 suggests the idea of an elliptic annular system seen very obliquely. A series of very elliptic shells enveloping the nucleus, seen somewhat obliquely, would perhaps also present the same aspect. The dark chink in Plate XXXVII. fig. 8 might indicate either a real opening, the system being an elliptic ring, or merely a line of comparative darkness, the section through the axis of a very long narrow elliptic shell. In Plate XXXVII. fig. 9 there is a well-marked stratification, which might possibly be the appearance, Plate XXXVII. fig. 7, on the first supposition, would present if seen in another direction. It is to be hoped that as observations multiply, and these extraordinary objects which abound in the heavens are seen in various directions, we shall gradually become acquainted with their real form. At present further conjectures would be to no purpose.

The remaining sketch, Plate XXXVIII. fig. 17, is the dumb-bell nebula as seen with the 6-feet telescope: the sketch is by Mr. JOHNSTONE STONEY, and the form of the nebulosity and its various gradations of intensity have been represented with considerable fidelity. There was no subsequent opportunity of marking in the stars, and therefore they have been inserted at random to complete the general effect, and many minute details are still wanting to make the figure complete.

As we have proceeded with our task of re-examining Sir JOHN HERSCHEL's Catalogue, several groups of nebulæ have been discovered, although new objects have not been as yet sought for. In some cases a nebulous connection has been detected between the individuals of the group, in others not. Sketches have been made and some measures taken. The whole subject of the grouped or knotted nebulæ is one of deep interest; but we have not proceeded sufficiently far with it to make it worth while to enter upon it in the present paper, and it only remains to point out a defect common to all the sketches which might mislead if not specially noticed. In sketching we necessarily employ the smallest amount of light possible, very feeble lamp-light, especially where the objects or their details are of the last degree of faintness. To see the sketch as we proceed it is often necessary to mark it too strongly: this would be of little moment if the excess of colour was always in the same proportion, especially as different eyes form a very different estimate of the relative intensities of a nebula and its representation on paper, but it is not so; the contrast between the faint and bright nebulæ and between the faint and bright parts of the same nebula is very liable to be made too slight. The most important error to guard against is that of supposing that the well-marked confines of the nebula on paper really represent the boundaries of the object in space in all cases. Frequently there is a very gradual fading away at the edge, the last trace of which is either a luminous mist becoming rarer till imperceptible; a gauge-like tissue of the faintish imaginary flocculi, or hairy filaments, which become finer and more scattered till they cease to be visible, showing that the real boundary has not been seen, and that the form of the object would alter if additional optical power could be brought to bear upon it. The same remark applies to the faint interior details, in most cases probably only in part seen.

Plate XXXV. figs. 1 and 2 are seen on a scale of half an inch to a minute; the others are on no regular scale: they are about the size of the figures which accompany Sir JOHN HERSCHEL's Catalogue, the smaller however have been somewhat enlarged where there were details which otherwise could not have been well represented.

Annexed are a few remarks relating to each figure, which seem to make the information conveyed by it more complete: they are for the most part extracts selected from our journal of observations; in a few cases, however, to save space, merely the substance is given.

Where the 3-feet instrument was employed it is specially mentioned; in every other case it was the 6-feet instrument.

Plate XXXV. fig. 1, H. 1622.—This object has been observed twenty-eight times with the 6-feet instrument; it had been repeatedly observed previously with the 3-feet instrument.

September 18, 1843.—Observed with the 3-feet instrument; power single lens.

I—I

1-inch focus; a great number of stars clearly visible in it, still HERSCHEL's rings not apparent, at least no such uniformity as he represents in his drawing.

April 11, 1844.—Observed with the 3-feet instrument, two friends assisting; both saw centre clearly resolved.

April 26, 1848.—6-feet instrument. Saw the spirality of the principal nucleus very plainly; saw also spiral arrangement in the smaller nucleus.

The following measurements were taken by my assistant, Mr. JOHNSTONE STONEY, in the spring of 1849 and 1850.

	Mean of the observations of position.	No. of observations.	Greatest difference between observations and the mean.	Mean of the observations of distance.	No. of observations.	Greatest difference between observations and the mean.
N. n.	16 34	4	3 27	4 22·2	4	9·6
N. 1.	52 4	1	......	2 6·6	1	
N. 2.	54 0	4	1 57	5 0·0	4	5·4
N. 3.	104 20	2	2 3	2 45·6	2	3·6
N. 4.	111 57	2	0 40	4 3·6	2	0·6
N. 5.	165 35	2	0 31	1 43·2	2	1·1
N. 6.	191 42	1	......	3 54·0	1	
N. 7.	211 2	1	......	2 36·6	1	
7, 8.	270 42	1	......	0 34·8	1	
N. 9.	231 32	4	3 35	1 23·4	3	6·6
9, 10.	197 57	1	......	0 27·0	1	
N. 11.	279 21	4	4 18	1 49·8	3	22·2
11, 12.	225 27	1	......	0 12·6	1	
N. 13.	281 37	2	0 22	3 59·0	1	
14, 15.	297 15	1				
N. 15.	310 34	4	4 17	2 55·8	4	13·8
N. α				3 22·8	2	0·1
N. β	5 7	...	......	1 28·2	3	3·0
N. γ				2 37·8	3	2·4
N. δ				1 46·2	1	
N. ε				2 46·8	1	
N. ζ	95 7	...	......	1 40·8	1	
N. η				3 15·6	1	

Observations.—There is a great discrepancy between the measured position of 11 and 12 and the rough diagram made at the time of observation.

N. 13 is twice noticed in the observing-book.

Once N. 11, 13 is taken as one position; the other times N. 11 and 13 are taken separately, N. 13 being made 1° 40′ less than N. 11; hence 270° 31′ is a more probable position for N. 13 than that given in the Table.

The Greek letters are perpendiculars from N. on tangents to the outsides of the convolutions, the tangents from α, β, γ being vertical, that is, parallel to the position 95° 7′, and those for δ, ε, ζ, η horizontal, i. e. parallel to position 5° 7′.

The greater part of the observations were made when the eye was affected by lamplight, which made it difficult to estimate correctly the centre of the nucleus; it was of importance that no time should be unnecessarily spent, and after the lamp had been used a new measure was taken, as it was judged that the object was sufficiently seen. With the brighter stars this would frequently happen before the nucleus was well defined, as all impediments to vision seem to affect nebulæ much more than stars the light of which would be estimated as of the same intensity.   In the foregoing list the greatest discrepancies are in the measures of bright objects, and this is probably the proper account of it.   No stars have been inserted in the sketch which are not in the table of measurements.   The general appearance of the object would have been better given if the minute stars had been put in from the eye-sketch, but it would have created confusion.

Plate XXXV. fig. 2, H. 1173.—This nebula has been repeatedly observed with the 6-feet instrument.

March 11, 1848.—Spiral with a bright star above; a thin portion of the nebula reaches across this star and some distance past it.   Principal spiral at the bottom, and turning towards the right.

March 20, 1848.—Spirality very evident, though night bad: nebula not traced to upper star.

April 16, 1849.—Took measures of the stars 1, 2.

April 17, 1849.—Took measures of the stars 1, 2, 3, 4 from the nucleus; they are as follows :—

No.	Mean of observations of position from north in direction *n. f. s. p.*	No. of observations.	Greatest difference between mean and observation.	Mean of observations of distance.	No. of observations.	Greatest difference between mean and observation.
1.	34° 1′	1	°	2′ 54·6″	2	9·6″
2.	80 35	2	0 18	1 46·3	3	14·4
3.	117 3	3	0 23	1 48·4	4	13·6
4.	177 57	1	......	2 48·1	1	

Three very minute stars in the eye-sketch have not been inserted, not having been measured.

Plate XXXVI. fig. 3, H. 604.—This nebula was observed frequently with the 3-feet instrument, but nothing remarkable seems to have been made out, except the resolvable character of the nucleus.   It was first observed with the great telescope, March 24, 1846, and a tendency to an annular or spiral arrangement discovered; night bad; March 5, 1848, sketched.

March 9, 1848.—" Night excellent, a spiral seen in an oblique direction, resolved well, particularly towards the centre, where it is very bright; Dr. ROBINSON observing."   Observed March 3, 1850; badly seen.

With the single exception of March 3, 1850, we have unfortunately no recent observation of this extraordinary object: it has been passed over, because to observe it, except on a very fine night, would be waste of time.

Plate XXXVI. fig. 4, H. 2205.—Observed frequently, and by many friends.   The drawing represents the object with considerable accuracy.

" September 10, 1849.—Spiral, but query whether this is not more properly an annular than a spiral nebula."

The details are faint, but can be seen on any moderately fine night.

Plate XXXVI. fig. 5, H. 131.—This figure represents the central portion of a very large nebula. The nebula itself has not been sufficiently examined, but as yet no other portion appears to have a spiral, or indeed any regular arrangement. The sketch is not very accurate, but represents sufficiently well the general character of the central portion.

"September 6, 1849.—A spiral.

"September 16, 1849.—New spiral; α the brightest branch; γ faint; δ short but pretty bright; β pretty distinct; ε but suspected; the whole involved in faint nebula, which probably extends past several knots which lie about it in different directions. Faint nebula seems to extend very far following: drawing taken.

"September 10, 1849.—An attempt at a drawing taken: fog.

"October 1849.—The whole nebula in flocculi."

Plate XXXVII. fig. 6, H. 444.—"December 19, 1848.—Bright star between; tails and curved filaments; perhaps annulus around the two nebulæ.

"December 22, 1848.—Sketch made.

"February 11, 1849.—Lower streak seems to reach the filaments of right-hand nucleus."

Plate XXXVII. fig. 7, H. 854.—"March 31, 1848.—A curious nebula with a bright nucleus; resolvable; a spiral or annular arrangement about it; no other portion of the nebula resolved. Observed April 1, 1848, and April 3, with the same results."

Plate XXXVII. fig. 8, H. 1909.—"April 27, 1848.—A very bright resolvable nebula, but none of the component stars to be seen distinctly even with a power of a thousand. A perfectly straight and longitudinal division in the direction of the major axis. Resolvability most strongly indicated towards the nucleus.

"May 2, 1848.—Not seen so well as on April 27. Darkness in the middle, along the major axis barely visible.

"April 1849.—A long ray elliptical. Major axis perhaps eight times minor axis. Surface somewhat broken up, and a slight darkness in the direction of the major axis: night indifferent: at intervals a few stars faintly perceptible."

Plate XXXVII. fig. 9, H. 1397.—This sketch was made with great care by my assistant, Mr. JOHNSTONE STONEY, and I have no doubt it is very accurate. Observed and sketched, April 19, 1849. It had been previously observed, March 26, 1848, by my former assistant, Mr. RAMBAUT, and I find the following note by him:—"A most extraordinary object, masses of light appear through it in knots."

Plate XXXVII. fig. 10, H. 399.—Observed December 22, 1848, February 11, 1849, and January 16, 1850, when the drawing was taken. The two comparatively dark spaces, one near the vertex and the other near the base of the cone, are very remarkable.

Plate XXXVII. fig. 11, H. 838.—September 27, 1843.—(3-feet telescope.) Night pretty good; a star in the centre and apparently ragged outline.

March 7, 1848.—(6-feet telescope).—Night bad : aurora. Darkness in the centre ; star not certainly seen ; outline ragged.

March 11, 1848.—Seen by Dr. ROBINSON and my former assistant, Mr. RAMBAUT ; sketch made of it. "Two stars considerably apart in the central region ; dark penumbra around each spiral arrangement, with stars as apparent centres of attraction ; stars sparkling in it, resolvable ; night excellent." Note by Mr. RAMBAUT : " March 5, 1848.—Saw two dark and very large spots in the middle ; Lord ROSSE remarked that all round its edge the sky appeared darker than the average."

" March 11, 1848.—Remarkably fine night ; a brilliant star in the centre ; also star to the right ; round each a black space (see sketch)." Note by Mr. RAMBAUT : " March 25, 1848.—Air steady, but slight haze ; large star visible. Only at one clear interval could I get a glimpse of the spiral arrangement of this nebula, which I should have totally overlooked had I not seen it so plainly on a former occasion.

" March 26, 1848.—Second bright star visible ; spiral arrangement hardly perceptible ; not seen so well as on the 11th of March.

" March 27, 1848.—Not seen so well as last night ; second star seen at rare intervals, power 468.

" March 28, 1848.—Night hazy, could not see second star."

" March 31, 1848.—Caught one glimpse of second star, but saw the large star very plainly.

" April 1, 1848.—Night hazy ; spiral arrangement little more than suspected ; nebula very faint.

" April 3, 1848.—Small star distinctly seen ; spirals tolerably well brought out ; hazy, but air steady.

" April 6, 1848.—First star seen easily, though hazy ; the second only occasionally ; spiral arrangement hardly discernible.

" January 1850.—Seen very imperfectly ; only one of the stars seen.

" March 9, 1850.—Second star only seen for a moment." Several attempts were made to procure measures of position and distance of the two stars this spring, but in vain, the season was so unfavourable. In 1848, the micrometer requiring illumination, no attempt was made. With the micrometer as at present mounted there would not have been the slightest difficulty in procuring measures.

Fig. 12, H. 464.—" Annular nebula at the edge of the cluster M. 46. Sketched December 22, 1848 annular, two stars in it.

" January 27, 1849.—A third star suspected in brightest part.

" January 29, 1849.—Third star strongly suspected.

" February 13, 1849.—Observed, nothing further.

" March 16, 1849.—Saw but two stars in it."

Fig. 13, H. 2241.—" October 31, 1848.—Has a central spot, at moments very dark.

" December 13, 1848.—Nothing more, except perhaps that faint external annulus extends further than had been seen before.

" December 14, 1848.—Note by Mr. JOHNSTONE STONEY:—' Three stars near it, somewhat in this fashion; showed it to Sir JAMES SOUTH.'

" December 16, 1848.—Sketches made by Lord ROSSE and Mr. JOHNSTONE STONEY.

" December 19, 1848.—Drawing confirmed."

Fig. H. 14, 2098.—" Observed October 23, 1848, and sketch made.

" October 25, 1848.—Sketch confirmed.

" August 16, 1849.—Position of ring taken with an eye-piece furnished with a level and a position circle.   Inclination of ring to horizon 9°."

Fig. 15, H. 450.—" February 20, 1849.—Most astonishing.   The star perhaps a little nearer the np edge.   Drawing made; breadth of ring less on f side.

" February 22, 1849.—Observed again; dark space to the right of star.

" January 16, 1850.—Observed; examined with the 700 and 900 eye-pieces; both the dark and the bright rings seemed unequal in breadth; the light appeared unsteady and flickering.   The night was rather foggy, but the sky black."

Fig. 16, H. 361.—" January 28, 1849.—ı Orionis.   Dark space in the nebula containing nearest companion; light nearly equable; sketch made; 3-feet telescope employed.   All the stars in the neighbourhood are nebulous, of these two a little sp, last seem to have dark spaces as in figure.   To the nf of this there is another smaller double star, suspected to have similar dark spaces to ı Orionis.

" February 16, 1849.—Three-feet telescope confirmed observation of January 28, 1849.

" March 17, 1849.—Large triple star to south of nebula Orionis; confirmed observation of opening in its atmosphere, also the openings at double star sp last."

Fig. 17, H. 2060.—" Observed September 9, 1849.   Drawing commenced.

" September 16, 1849.—Drawing proceeded with; examined also with 3-feet telescope to find if any evidence of change since drawing in Philosophical Transactions was taken; none decisive."

List of some remarkable Nebulæ.

Spiral or curvilinear.

H. 142, 262, 327, 695, 749, 910, 1002, 1211, 1312, 1368, 1451, 1570, 1776, 2172.

With dark spaces.

264, 368, 491, 514, 692, 731, 788, 857, 887, 1107, 1225, 1909, 2241.

Ray with split.

1041, 1149, 1357.

Knotted nebulæ.

84, 257, 320, 409, 446, 581, 1274, 1901.

XXVIII. *On the Construction of Specula of Six-feet Aperture; and a selection from the Observations of Nebulæ made with them. By the Earl of* ROSSE, *K.P., &c., F.R.S.*

Received June 5,—Read June 20, 1861.

THE period seems now to have arrived when it may be proper to lay before the Royal Society some further account of researches in sidereal astronomy, carried on with a Newtonian telescope of 6-feet clear aperture.

The observations extend over a period of about seven years, during which few favourable opportunities were lost; still in our climate, where there is so much cloudy weather, a year's work, measured by the number of hours when nebulæ can be effectively observed, is not considerable. Here in winter the finest definition we have, and the blackest sky, is usually before eleven o'clock, after which the sky becomes luminous, and the fainter details of nebulæ disappear. In spring and autumn the change is neither so early nor so decided; but the nights are shorter. Guided by Sir JOHN HERSCHEL's admirable Catalogue, we have examined almost all the brighter known nebulæ except a few in the neighbourhood of the pole, and a great proportion of the fainter nebulæ. No search has been made for new nebulæ; very many, however, have been found accidentally in the immediate neighbourhood of known nebulæ, but for the most part they were faint objects presenting no features of interest. In every case where any peculiarity was detected, as for instance the convolution of a spiral, dark lines, or dark spaces, a rough sketch was made, and the more remarkable objects were selected for examination on favourable nights, when the details were carefully filled in, sometimes with the aid of the micrometer. The very faint objects, and even the brighter, where there was a simple gradation of colour and no peculiarity of form, after having been examined on a tolerably good night, were rarely examined again. In our ever-varying climate, when we employ high powers and large apertures, vision is impeded more or less by the unsteadiness of the air; it is impeded also by haze; and in both respects the condition of the air varies immensely from night to night, and from hour to hour. The speculum also is not uniform in its action. With such sudden alternations of temperature, in a moist climate, it is frequently dewed, and gradually tarnishes. Artificially heating it would be a remedy; but it would be an objectionable one, and we have not employed it. From all these causes we can scarcely say that any one object has been examined under a combination of favourable circumstances; still it is not now probable that with the present instrument any remarkable additions will be made to the details of nebulæ already carefully sketched, except in very favourable states of the atmosphere. Occasionally the air is so transparent and so steady, that magnifying power may be pushed very far; and then, perhaps, something new comes out. Such opportunities, however, are rare;

4 z 2

and the progress made is necessarily so very slow, that I think it would be inexpedient longer to keep back this paper in the distant hope of making it in some respects more complete.

As to the instrument, a slight description of it has already been given in the 'Transactions' for 1850, but without details, and in the 'Transactions' for 1840 the process employed in the construction of specula of 3-feet aperture was fully explained; but in passing from specula of 3-feet aperture, and about twelve hundredweight, to specula of 6-feet aperture and four tons, although the same principles were our guide, difficulties were encountered which called for new contrivances and additional precautions. It will, I think, be useful to give a short account of the process by which the 6-feet specula were made, and some details as to the mounting, supplying at the same time the best answers we can to the questions, so often put, What really are the optical powers of the instrument? What are the merits and demerits of its form of mounting, after an experience of more than ten years? Would it be possible to construct a larger one, and, if so, would there be anything gained? As there seems to be a desire to employ large instruments in different parts of the world, would it be possible to lay down instructions sufficiently precise to enable a mechanical engineer, without a previous apprenticeship, to undertake the construction of large instruments as a matter of business?

About one ton and a quarter of speculum-metal can be melted in one crucible; and up to that weight there is no difficulty whatever in casting a speculum, and the instructions in the 'Transactions' for 1840 are amply sufficient to enable any engineer to do so. Each time, however, the crucible increases in circumference from the pressure of the metal, and after seven meltings we found the increase to amount to 4 inches. One ton and a quarter is therefore about the limit for a separate melting, and for larger specula we must employ several crucibles. The tin and copper must be previously combined in smaller crucibles, holding not more than three or four hundredweight, as the heat required is much greater than in the second melting. Three crucibles were employed in casting the 6-feet specula; and we proceeded thus:—

The crucibles (which had been cast by Messrs. DEWER of London, with the precautions detailed in my former paper) were placed in three separate air-furnaces upon cast-iron stands about 8 inches deep, and of somewhat larger diameter than the crucibles, to protect them from the immediate current of air passing through the fire-bars. A brick pillar from the bottom of the ash-pit relieved the furnace-bars from the weight they would otherwise have had to sustain. The furnaces were round, 4 feet diameter, and 6 feet deep to fire-grate,—constructed as an ordinary air-furnace, with a door at the ash-pit to regulate the admission of air. The three furnaces were worked by one stack. In heating a crucible, it is necessary that the temperature should be raised gradually, *beginning at the mouth*; otherwise it will be very liable to crack. To satisfy this condition, the crucibles (of course empty) having been placed on their supports, and the furnaces filled with good peat, the fires were lighted at the top; and in about *ten hours* the crucibles were of a proper temperature for the reception of the speculum-metal, which of course was introduced gradually.   In about twenty-six hours from the time the fires were lighted the metal was ready for pouring.   Peat of good quality is about equal to wood in heating-power, when consumed in furnaces where there can be no accumulation of charcoal.   The mould was constructed on the principles explained in my former paper; but, the scale being now so much enlarged, little matters of detail, which might have been before overlooked with impunity, were found to be of vital importance. The bed of hoop-iron was 6 feet 6 inches diameter, and 4 inches thick.   We had not at the time a lathe sufficiently large to turn it; and therefore it was turned horizontally, on the machine which was to grind and polish the future speculum.   To remove little irregularities arising from the imperfection of the turning-apparatus, the bed of hoops was ground for two or three days with a disc about 6 feet diameter, composed of fragments of sandstone cemented together within an iron ring.   The annealing-oven was built on four arches communicating with two low chimneys.   The floor being laid upon the arches, could easily be heated to redness.   The interior of the oven was 8 feet by 10. For want of room, the brickwork at the ends was but 2 feet thick, the sides nearly 4. The thrust of the arch was, in the usual way, sustained by bolts.   The crucibles were raised from the furnaces by a crane and tongs just as at the Mint, and placed in rings swinging on trunnions a little above the centre of gravity of the mass.   The metal being of a proper temperature, levers were fixed upon the trunnions, and at a signal the crucibles were simultaneously inverted as rapidly as possible.   The operation of pouring was accomplished in about three seconds.   If the metal was not poured rapidly, the conducting-power of the iron surface is so great that partial solidification would take place, and the casting would be imperfect.   In about twenty minutes the metal was solid throughout; the frame containing the sand forming the sides of the mould was then removed, and the speculum, being grasped by an iron ring, was drawn into its place by a capstan.   The temperature of the oven was red, just perceptible in the dark, about 900°. All the apertures were then closed; and in about six weeks the speculum was cool. When removed from the oven the speculum was found perfect; but the radius of curvature was much longer than it should have been, which rendered the grinding a very tedious operation.   The cause, however, was obvious; the floor of the oven had been laid carefully flat to prevent warping; no other precaution had been taken; indeed, no other had been necessary with the 3-feet specula.

The speculum was removed from the oven to the bed of supporting levers in the following manner:—A pit was dug about 4 feet deep, near the oven, commanded by a crane.   The speculum, weighing about four tons, was drawn out of the oven in the same way that it had been drawn into it.   Planks were provided for the speculum to slide upon to the edge of the pit, into which it was lowered gently, the ring still grasping it. The speculum was now resting principally upon its edge, the face supported by the side of the pit.   By means of *wooden* handspikes, and with little effort, the speculum was made to rest entirely on its edge, bearing upon the soft earth.   Two bars, 7 feet long each, and 2 inches square, one of them cranked in the centre, were placed against the back in the shape of a cross. To prevent metallic contact, the bars had been bound round with woollen cloth. Strong planks were placed against the face, and screw bolts were passed through the planks and projecting ends of the bars. The speculum was thus encased, and was easily raised by the crane *face up*. In the mean time a strong wooden platform had been made with three iron pillars securely fixed in it, about 2 feet long each, and so disposed as to support the speculum with the least strain. The frame carrying the supporting levers, to be hereafter described (Plate XXIV. fig. 1), was placed upon this platform, the three iron pillars passing through interstices in the levers; and the speculum was lowered till it rested upon the pillars, the levers being considerably below it. The bars and planks encasing the speculum were then removed, and the frame and levers raised by the crane till the speculum was completely supported by them. It now rested on its levers, and was taken to the grinding-machine. I have been thus minute in describing the means we had employed in removing the speculum from the oven, turning it over, and placing it on its bed of levers, as in the arts they have never to deal with a material at once so heavy and so brittle; and we were guided by long experience, which others may not have had.

This speculum had been more than a month upon the grinding-machine, and was just ready to be polished, when it was broken by an accident. Immediate preparations were made for recasting it. While the speculum had been in the annealing-oven we had finished a powerful lathe for turning the grinding-tools, with a slide-rest moving in the proper curve. The bed of hoops was placed upon that lathe, and its radius of curvature adjusted: the floor of the oven also was cut roughly to the same curve. As we were anxious to guard, as far as possible, against contingencies, and to secure a working speculum with the least delay, we were satisfied to employ an alloy somewhat lower than on the former occasion, and an ingot of speculum-metal was added which contained more than the proper proportion of copper. A little additional copper diminishes the brittleness considerably, while it increases the liability to tarnish.

The speculum was successfully cast, but the surface was covered with minute fissures, about the breadth of a horse-hair. These we resolved to grind out. The grinding was very tedious, partly owing to the metal being a little below standard, and partly to the deepness of the fissures. After the first day's grinding, the fissures, which previously were scarcely perceptible, became much enlarged, owing to the edges chipping away; and the whole surface thus became, as it were, covered with large wrinkles. The process of abrasion is necessarily extremely slow, as both the velocity and the pressure are kept within very narrow limits, to prevent the evolution of heat, which would crack the speculum. The grinding continued for nearly two months, the machinery working for part of the time at night; and a few of the fissures were so deep that even then the traces of them were perceptible. The speculum was then polished; and its performance fully equalled our expectations.

A telescope intended to be constantly employed requires two specula. We had now leisure to encounter delays and difficulties in endeavouring to procure a second speculum free from the defects of the one already finished. We had satisfied ourselves that the fissures were owing to our having employed the bed of hoop-iron in the state in which it was when taken from the turning-lathe. The surface, though nicely turned, was not as smooth as the surface of a solid disc would have been: a slight yielding at the edges of the hoop-iron, and a slight spreading under the pressure of the tool, had produced little irregularities; and although the surface had been carefully "black-washed," the speculum-metal had encountered too much friction in the act of contracting after it had become nearly solid, and thus had been filled with superficial rents. On the first occasion there had been no fissures, but the bed of hoop-iron had been ground; the remedy was therefore obvious.

The third speculum was successfully cast; but on opening a small aperture, and looking into the oven before it was quite cold, it was observed that the speculum was cracked through the middle. The temperature of the speculum was found not to be quite uniform; and that circumstance, taken in connexion with the direction of the crack, seemed to point out the cause: the ends of the oven, from want of room, had been made thinner than the sides. The first speculum had probably been strained by the same cause, and rendered more fragile.

The oven being ready, an attempt was made to cast a fourth speculum, which failed. We had each time, before the bed of hoop-iron became cold, saturated it with tallow to prevent the formation of rust between the hoops, which would have rendered the surface impervious to air; but just before it was again employed it was made red-hot, and the tallow burned out. On this occasion, by an oversight, the bed of hoop-iron had not been sufficiently heated, and there remained some of the tallow unconsumed, which, being vaporized in large quantities, produced an ebullition which made the casting as porous as pumice-stone. This speculum, of course, was not annealed, and the following day it was in small fragments.

The fifth speculum, being in every respect a perfect casting, without the slightest blemish and of a proper curvature, was ground and polished in about a month. It is desirable that the bed of hoop-iron when the metal is poured should be warm, so as to prevent the possible deposition of moisture; but if much hotter than this, it at once dries up the sand, and it is difficult to make the mould secure. In the whole of the operation I have described, one of the difficulties is to time each stage. If the mould was prepared too long, the bed of hoop-iron might become cold and damp; on the other hand, if the mould was not ready when required, it might be hazardous to keep the crucibles so long at the pouring-heat. It may, perhaps, be as well to add that the crucibles, when in the pouring-gimbals, require to be thoroughly skimmed, as particles of coal falling upon the hoop-iron would be immediately entangled in metal not rising to the top: the skimming should be done promptly, lest the metal fall below the proper temperature. Any considerable delay in drawing the speculum, when solid, into the annealing-oven would be fatal; therefore there should be ample capstan power to overcome the difficulty which usually is experienced in detaching the speculum from the mould.

The last speculum is but $3\frac{1}{2}$ tons, and is therefore considerably weaker than its predecessor; and by carefully comparing the two specula at low altitudes, we have been made thoroughly sensible of the great importance of strength in preventing flexure. There are little irregularities in the action of the supporting levers, which are much more injurious to the definition of the weaker speculum than the other; and although these irregularities may be susceptible of further diminution, I think there would still be sufficient gain to make it worth while to cast a third speculum considerably heavier than either of the others.

In the 'Philosophical Transactions' for 1840 I have endeavoured to explain the principle upon which the bed of hoop-iron acts; some, however, seem to have attributed larger effects to it than I have, and of a different kind. It has been supposed by some that a molecular change takes place, somewhat similar to that which has been observed in the case of very small portions of speculum-metal rapidly cooled, while by others the change has been compared to the " chilling of cast iron," to which I think it bears no analogy: cast iron when chilled becomes almost as hard as hardened steel; there is an exudation of graphite—in fact, a chemical change the exact nature of which seems to be imperfectly understood: there is no such change in speculum-metal, it becomes actually softer. To obtain sound castings, all which seems necessary is so to manage the process that solidification must begin at one surface and proceed regularly to the other. By employing the bed of hoop-iron the object is effected with certainty; but the engineer may employ other means, perhaps sufficient for the purpose, which, under varying local circumstances, may be cheaper and more convenient.

Possibly some useful hints may be gathered from a slight glance at the successive steps by which we obtained a clear view of the principle by which the founder should be guided in making large castings of speculum-metal.

About the year 1827, on commencing a series of experiments on speculum-metal, I procured a small flat speculum from Mr. TULLY, and two similar specula from Mr. CUTHBERT, as specimens of the art in its most advanced state. I also procured from Mr. CUTHBERT several small unwrought castings of about two ounces weight to practice upon. Mr. TULLY's specula were cast in the ordinary way in sand, and polished with rouge: but Mr. CUTHBERT's were cast in contact with iron, and so cooled instantaneously; they were polished with putty. All the specula for Mr. CUTHBERT's microscopes were made in a similar manner. He was under the impression that speculummetal cooled instantaneously was more suitable for his purpose than common speculummetal—that it was sounder, more compact, and resisted better the action of emery. These specula were accidentally exposed to the air of the laboratory for a considerable time; and at length we remarked that Mr. CUTHBERT's specula had somewhat lost their polish, while TULLY's speculum was as bright as ever. The inferiority of CUTHBERT's specula we attributed to an excess of copper, but with further experience we came to a different conclusion. We had several early samples of chilled speculum-metal, and corresponding samples of the same metal cooled gradually. They were obtained in this way:—In our experiments on the alloys of tin and copper, we were in the habit of taking out a sample after each addition of tin. When cool, a small piece was broken off the sample, and the fracture and colour examined; the remainder was then hastily ground and polished on a succession of revolving laps. The experiments were very numerous; and to save the time lost while the sample was cooling, we at first applied water cautiously, and then adopted the device of pouring the sample into a ring laid upon the face of an anvil. Samples of a few ounces weight frequently cracked upon the anvil, but with water they usually flew into many pieces. We soon, however, found that the attempt to save time by cooling the samples instantaneously was a step in the wrong direction, as it was only from samples cooled slowly that reliable information could be obtained as to the qualities of the future casting. There was, however, this result, that we at length came to the conclusion that instantaneous cooling was unfavourable to permanence of polish. In the progress of these experiments we also observed that the rods of speculum-metal formed in the air-holes of damp sand-moulds, also the thin plates formed at the junction of the upper and lower moulds, were of unusual strength. We were not then aware of the fact that alloys of tin and copper are softened by sudden cooling, which would have accounted for the liability to tarnish, and the great increase of strength.

Mr. POTTER, in Sir DAVID BREWSTER's 'Journal' for 1831, directs attention to the *apparent* hardness and soundness of speculum-metal cooled instantaneously; but he does not appear to have operated upon a larger scale than Mr. CUTHBERT, his castings not exceeding $1\frac{1}{2}$ ounce. Mr. M'CULLAGH seems also to have noticed the same facts; and, indeed, it is not likely they could have been unnoticed by any one who had been engaged much in speculum-casting; but the obvious fact that any considerable mass of an alloy with such large expansions and contractions as speculum-metal, and so brittle, must fall to pieces if cooled rapidly, would have forbidden the attempt to manufacture large telescopes with such a material.

In all our earlier experiments the castings were made in damp sand, precisely as in the common process of casting iron or brass. Where the founder, however, aims at the best results, especially in brass-casting, he dries the mould: he thus escapes the mischief sometimes arising from the evolution of hydrogen, which, unlike steam, makes its way through the sand with difficulty. Steam in small quantity does no mischief, because it enters the interstices of the sand, where it is immediately condensed.

In the hopes of better results we dried the moulds; but, strange to say, the castings were less perfect. At the low temperature at which specula are cast, the tin acts but very little on water, and there is no injurious evolution of hydrogen; therefore, in that respect, there was nothing gained by drying the mould, while we found, after a great number of specula had been broken up, that in dry sand the progress of solidification had been less regular than in damp sand, and that this was owing to the circumstance that in dry sand the solidification had commenced irregularly in all directions, while in damp sand the upper surface had remained longer fluid than the lower surface, especially where the specula were of considerable thickness. To explain this, it is only necessary to remark that, when metal enters the damp-sand mould, heat will be immediately abstracted from it; and as it rises in the mould by each successive addition of hot metal, it will somewhat dry the upper sand-surface before the metal reaches it. When, therefore, the mould is full, the lower surface will be cooler than the upper: this does not happen when the mould is dry. In casting very thin specula, there is no time for successive actions; the upper and lower surfaces solidify simultaneously, and there is a tendency to separation in an intermediate plane. In some cases the separation was so complete, that a slight concussion actually divided the speculum into two discs. In discs of brass there is often, from the same cause, a very thin plane of porous metal running through the centre; and where this occurs in the plate of an air-pump, a bouching is inserted to cut off the communication between the external air and the central aperture. Keeping these facts in view, it would be naturally expected that by employing very open sand, as damp as possible, for the lower surface of the mould, and dry sand for the remainder, the best results would be obtained: and such was the case; and where other means are not at hand, specula of 10 or 12 inches diameter can thus be easily obtained, provided they are of considerable thickness. This device, however, was not successful when we endeavoured to procure thin plates to face the compound speculum described in the 'Transactions' for 1840; the solidification of the upper and lower surface was too nearly simultaneous, and therefore there was irregular contraction: consequently a metallic surface was employed, from which the plate was removed the moment it was solid.

There is yet another method of procuring excellent specula of moderate dimensions, which was dismissed, perhaps, in too summary a manner in the account of experiments in the 'Transactions' for 1840. It has this to recommend it, that it can be carried out by persons who have had no experience in the management of melted metal; and it is desirable to smooth the way, as far as possible, for beginners, who may, perhaps, by early success, be induced to proceed further. A cast-iron mould can easily be made at any foundry; it must be at least two and a half times as deep as the required speculum; it is to be placed in a temporary air-furnace, resting, like a muffle, upon two very strong deep bars, and is to be made perfectly level. The grate should be made of moveable bars, which can be withdrawn at the conclusion of the process, so that the fire may fall into the ash-pit. If charcoal is employed, the draught will be sufficient to produce the necessary heat without a chimney. The proper quantity of speculum-metal, in pieces, is then introduced, and the cover put on. It is important that no pieces of charcoal should get in before the metal is melted, as they will often be found in the face of the speculum. As soon as the metal is melted, the cover is taken off, and the moveable bars are drawn out. The metal is then stirred with a broad flat tool, passed everywhere over the surface of the mould to detach air-bubbles, and without loss of time a jet of water is thrown against the bottom of the mould, through a rose with exceedingly small holes, and distributed evenly. The action of the water must be suspended the moment the temperature of the mould is reduced to a dark red, lest it should crack; but the operation should be repeated at intervals of a few seconds, to keep the reduction of temperature permanent.   As soon as the metal is solid at the surface, the furnace is to be closed up completely, and the speculum is thus annealed, the furnace acting as an annealing-oven.   The blocks of speculum-metal, which were sawn up into plates, as described in the 'Transactions' for 1840, were made in this way, excepting that air was employed instead of water.   A large hand-fan furnishes a sufficient blast; and when such an instrument is within reach, air is perhaps preferable to water, as it is more easily managed.   The cracking of the moulds (the difficulty we encountered in the experiments alluded to) we subsequently ascertained was owing to excess of water.   It is important that the temperature of the metal should not pass the melting-point, to prevent the large development of a crystalline structure.

Whether specula are cast according to the first process, when the moulds are of damp sand and solidification commences at the lower surface in the way I have explained—or by the second process, when the moulds are of iron and solidification also commences at the lower surface, owing to the action of some cooling medium, through the iron—or by the third process, when the same effect is produced by the exterior mass of iron which prevents the interior surface of the mould from attaining the temperature of fusing speculum-metal, the result is very similar: there is the same molecular arrangement more or less developed, and the fracture presents the same characters: the axes of the crystals are directed to the cooling surface, in obedience to the general law, stated by Mr. MALLET in the following words, that when the particles of crystalline solids* " consolidate under the influence of heat in motion, their crystals arrange and group themselves with their principal axes in lines perpendicular to the cooling or heating surfaces of the solid—that is, in the lines of direction of the heat-wave in motion, which is the direction of least pressure within the mass."

It is scarcely necessary to add that there is no resemblance in this molecular arrangement to that of a small speculum cooled instantaneously.

Enough has now been said to enable a skilful founder to follow the course which we pursued in casting specula of 6-feet aperture.   The principles are, in fact, the same which he must have had in view in executing works of unusual difficulty in cast iron: with speculum-metal, however, the difficulties are far greater, and therefore every part of the operation must be more rigorously governed by sound principles.

The photograph of the speculum on its supporting levers (Plate XXIV. fig. 1) will give perhaps all the information which may be required as to the general nature of the arrangement.   The ring in which the speculum is suspended was removed, as also some of the apparatus connected with the levers, to prevent confusion.   The diagram fig. 2 represents in plan the arrangement of the levers.   The cast-iron carriage, of about $1\frac{1}{2}$ ton weight, carries three ball-and-socket joints, directly under the centre of gravity of three equal sectors, into which the speculum may be supposed to be divided.   The

* MALLET 'On the Construction of Artillery,' p. 7.

5 A 2

K—1

centre of the ball is in the centre of gravity of the triangle, not merely as respects its plane, but thickness also. These three triangles, which we call primary, carry at their angles, by ball-and-socket joints, nine secondary triangles, supported at their respective centres of gravity: and they, in a similar way, carry twenty-seven tertiary triangles, each carrying three gun-metal balls of $1\frac{1}{2}$ inch diameter,—in all, eighty-one balls, which support twenty-seven equal portions of the speculum. Between the balls and the speculum twenty-seven thin brass plates are interposed, attached to the speculum by pitched cloth, not so much with a view of giving support between the balls, which would probably be quite unnecessary, but to make a smooth surface for the balls to roll upon without grinding the back of the speculum true. In each ball there is a small hole, and a thin brass wire is inserted in it, and secured with a wooden peg; this wire passes through a small hole in the lever, and is attached to a thin brass spring at the back, which yields as the ball rolls, and brings it back to its proper place whenever the ball is free from pressure. Without this contrivance, a very slight jerk, when the plane of the speculum is nearly vertical, would cause the balls to fall from their places. In practice, the motion of the balls is of course very slight.

It is evident that so long as the speculum is horizontal, equal portions are carried equably, and it is almost as free from strain as if it was floating on mercury. As soon, however, as we incline the speculum to the horizon, the lever apparatus does not act so perfectly. It will have been observed that the levers are not in the same plane; and this is disadvantageous in two ways: first, although the primary triangles balance in every position on the ball-and-socket joints, and therefore are indifferent as to position, the centres of the ball-and-socket joints carrying the secondary triangles are unavoidably above the plane of the centres of the primary supporting balls, and still more are the ball-and-socket joints carrying the tertiary triangles; consequently the secondary and tertiary triangles, by their weight, exert a force tending to make the planes of the primary triangles rotate in a vertical plane, and so disturb the equilibrium. The tertiary triangles in a similar manner, but to a much less injurious extent, act upon the secondary triangles. The lever apparatus, deducting the primary triangles, weighs about 600 lbs.; were it lighter it would not have the necessary solidity; and the disturbing action of the weight is so considerable that, when not counterbalanced by subsidiary contrivances, the action of the speculum at low altitudes is much impaired by it. The contrivance we employ is a system of levers, connected by wires with the ball-and-socket joints which support the secondary triangles, and acting at right angles to the plane of the speculum. The primary triangles, thus relieved from all lateral strain, are in a condition to do their duty effectively; and that seems to be sufficient in practice. Of course another set of levers might be attached to the ball-and-socket joints supporting the tertiary triangles; and then the whole system would be perfectly counterpoised in every position; that, however, seems to be scarcely necessary. It is evident that, were it not for the balls interposed between the levers and the speculum, any lateral motion of the speculum would introduce a disturbing force which would destroy the equilibrium. Lateral motion, however, must always exist in the different positions of the telescope, owing to the elasticity of materials; and it must act injuriously in some degree, in proportion to the force required to move the speculum on the balls. Great care has been taken to make the fittings of the ring in which the speculum hangs as perfect as possible, and to connect its joints and bearings with the iron carriage, so as to reduce the lateral motion of the speculum to the smallest possible quantity. We have not tried Mr. LASSEL's ingenious arrangement for relieving the edge pressure. Unless there were holes half through the speculum the experiment could not be fairly tried. Our 3-feet specula are also suspended in a ring, and are supported on fewer points by the lever apparatus, which, at a slight sacrifice of theoretical accuracy, has been thrown into one plane. We have rarely perceived any flexure of importance, except where the action of the levers had been impeded by rust; but the 3-feet specula, which weigh about thirteen hundredweight, are very much stiffer than the 6-feet specula, as is obvious on common mechanical principles. Upon the whole, I am inclined to think there is a better prospect of improving the definition of very large specula by increasing the original stiffness, than by endeavouring still further to eliminate slight disturbing forces.

The 6-feet specula were ground with an iron tool, divided into squares, precisely as the 3-feet specula. The squares were larger, about 2 inches each side, and were not formed by cutting but by casting. A tool cast from the same pattern was employed as a polisher, but the surface of the squares was cut up by turning, so as to leave no more than half an inch of continuous surface. The tools weigh about one ton one hundredweight each, and the iron is so disposed in them as to produce the utmost amount of stiffness. The photograph supplies all details (Plate XXIV. figs. 3 & 4).

In grinding, about two-thirds of the weight is at first taken off by a counterpoise acting through a system of levers attached to it in thirty-six points, on the same principle as the levers which support the speculum. As the process proceeds the contact becomes more general, the friction increases, and there is more heat developed; therefore the counterpoise is increased, till towards the conclusion the unbalanced weight of the grinder is reduced to about two hundredweight. Notwithstanding the great strength of the tool, we found that if after the grinding was over it was suspended by its centre, the flexure, after a week or two, became so great that on replacing it on the speculum and regrinding with it, the action commenced at the edge; it is therefore always, when not in use, suspended by its levers. The curvature of this tool was adjusted in the ordinary way by gauges. These, as they were to be employed in the adjustment of the speculum to focus, were made with great care. One side of each gauge was first made into a straight edge, by the well-known scraping process of Mr. WHITWORTH, and the two were then very slightly ground for a few minutes with fine emery to remove the marks of the scraper, but no more. Ordinates were then set out, an inch apart, and marked to the calculated length by means of an instrument applied to the straight edge, very similar to the joiner's gauge, but made of brass, with a fine scale and vernier. Through the extremities of the ordinates the curve was traced by means of a steel point, guided by a curved rule about 6 inches long. A pair of these little rules were made from calculated ordinates, and ground together. In adjusting the gauges to the curve so traced, nothing was employed but the file and scraper. The gauges were then slightly ground together with the finest emery and in very small quantity, and care was taken to distribute it evenly with a camel's-hair pencil. The grinding-tool was from time to time adjusted roughly to the curvature on a turning-lathe, which was accomplished with great facility, as the slide-rest was governed by a guide of the same curvature as the gauge, and there were adjusting-screws in the face-plate, by which the tool was made to run perfectly true each time it was replaced. When the curvature of the speculum was nearly exact, the remaining little changes in the radius of the tool were made with the file. We had no means of optically measuring the focus of the 6-feet mirrors while on the engine; therefore further precautions were taken. A brass wedge was made, about 3 inches long, at one end $\frac{1}{500}$th less than the vers sine of the circle of curvature, and at the other end $\frac{1}{500}$th greater. This wedge was cut into three parts, equal, greater, and less than the vers sine. A straight edge was then laid upon the speculum; and it was considered perfect when one of these just touched the straight edge, another passed under it without touching, and the third did not pass at all. So accurately was the adjustment as to focus made in this way, that neither of the specula differed from the calculated focus more than $1\frac{1}{2}$ inch.

The machinery is precisely on the same principle as that which we employ in working 3-feet specula, already described in the 'Transactions.' Instead of belts there are chains, working in V grooves. The driving-wheels are of wood, in several layers, the grain being disposed radially. The chains are made tight by straining-pulleys, and act perfectly. This species of machinery is neither compact nor elegant: when originally designed, its principal recommendation was facility of construction and facility of alteration, both important qualities where it was doubtful whether machinery unguided by hand, acting independently of the sense of touch, would answer at all. Some are surprised that machinery so rude should be employed, and successfully, in a mechanical operation where the utmost precision is required, a precision almost fabulous, and they compare it with the beautiful machinery in the mills where textile fabrics are made: but in figuring specula everything depends upon the principle; and so long as certain motions are communicated to the tool and speculum, machinery can do no more. The tool is raised and lowered by a screw passing through a carriage which moves upon a railway over-head upon the principle of the travelling crane, and the same mechanical arrangement removes the speculum with its lever supports from the grinding-machine to the truck upon which it is conveyed to the telescope. The screw is obviously the best mechanical power to employ, as its action begins and ends slowly, and there is therefore less danger of breaking the speculum. As a further precaution in raising or lowering the grinding-tool, thin wooden wedges between the tool and speculum are gradually introduced and withdrawn. In the final adjustment of the speculum to focus, the operation is much facilitated by a judicious management of the second eccentric; small variations in the radius of curvature are thus produced with great facility: where, however, there has been a considerable departure from the length of stroke necessary to produce a spherical figure, the speculum requires to be ground for twelve hours, or perhaps much longer, with the proper motions, before it is fit to be polished. Mr. WHITWORTH is of opinion that greater general accuracy of surface is obtainable by scraping than by grinding: the late Mr. A. Ross, as high an authority as any one in everything relating to practical optics, held very nearly the same language*. He attributed the defective action of the grinding-process to the unequal distribution of the grinding-powder, which, accumulating at the centre by capillary attraction, and at the edges by mechanical action, unduly shortened the radius of curvature at the centre and lengthened it at the edge. He employed the grinding-powder dry in producing flat glass surfaces, and believed he thus obtained a better result. It appears to me that by cutting up one of the surfaces into minute squares, in the way we have so long practised, the causes of unequal action are eliminated. The subject is a very important one, as there appears to be no other probable means of working solid materials into accurate surfaces for optical purposes than by some modification of the ordinary process of grinding and polishing. In the 'Philosophical Transactions' for 1840, there is a sketch of a grinding-tool such as we employed in the construction of 3-feet specula ; but I have scarcely noticed experiments on grinding, passing at once to the more important subject (as it appeared to me at the time), that of polishing. Something useful, however, may perhaps be extracted from our record of very numerous experiments on grinding plane and curved surfaces. In my very early experiments, the ordinary process for procuring a true plane by grinding three planes in alternate pairs was often repeated. Till we adopted the expedient of cutting up two of the surfaces into minute squares, our success was very limited. That device apparently removed all the difficulty, and we were then enabled to make large flat mirrors which bore the optical test well in *every part*, which was not the case before. When one of the surfaces is divided into minute portions, with sufficiently large and deep intervals, there can be no capillary action such as that described by Mr. A. Ross; neither can there be an accumulation of grinding-powder anywhere, because an immediate escape for it is provided; and if the grinding-powder is employed in very small quantities, and no addition is made to it for three or four hours before the termination of the process, there will be a high degree of comminution, and only just a sufficient number of minute particles to keep up the abrading action, probably nowhere more than a single layer. We may form some idea of the accuracy thus attainable by examining with a microscope the particles of emery so comminuted. No parts can be acted upon strongly except where they deviate from a spherical surface; too violent contact then, and consequently a destructive action, is prevented by the moisture interposed. Under such circumstances, with unyielding surfaces, *time* obviously cannot enter as a disturbing element, because there is no abrasion when there is no close contact.

* HOLTZAPFFEL, Mechanical Manipulation, vol. iii. p. 1229.

K—2

The principle of Mr. WHITWORTH's method may obviously be carried out, with glass or speculum-metal, by employing small laps and grinding-powders instead of the scraper; but as a scraped surface consists of a maze of curves of varying flexure, a surface ground in detail must always, I should think, in some degree partake of the same character, and, though it may not anywhere deviate much in general outline from the required form, minute deviations must exist in every part. M. FOUCAULT seems to have been successful in improving surfaces of moderate dimensions, by his ingenious process of testing and polishing in detail; how far such a process will succeed in improving large surfaces which have been in the first instance properly wrought, has not, as far as I am aware of, been ascertained. Our practice always has been to repolish when the surface, tested by the method described in the 'Transactions' for 1840, has proved to be defective. If a few glaring defects are at once seen, the whole surface is always faulty, though in a less degree.

The only change we have made in the polishing-machinery consists in substituting an elliptic for a circular wheel in driving the second eccentric. The major axis is at right angles to the throw of the eccentric, and is to the minor axis as three to one. The band is merely a rope working in a deep groove; and a straining pulley, freely acted upon by a weight, secures the necessary tightness in all positions of the ellipse. The obvious effect of this arrangement is to diminish the time the polisher overhangs the speculum, and so to reduce, to some extent at least, a source of error. We now employ in every case a separate tool for grinding and polishing, which is a great convenience, especially as we always regrind the speculum after it has been brought in from the telescope. There seems to be no doubt that in some cases considerable change of figure had taken place. The grinding-tool, when true, will be bronzed all over, and the speculum, when examined in every position as to light, will appear uniform.

We still consider the process of polishing described in detail in the 'Transactions' for 1840 as the best, with this addition, that we employ a combination of brown soap and ammonia, instead of pure water, during the latter part of the operation. We had then tried this lubricating mixture but too recently to feel justified in recommending it. The great objection, however, to the whole process is the difficulty of carrying it out. I have had communications from time to time from many persons in whose hands it has failed; and I am not surprised; for although everything has usually gone on smoothly when we were in the midst of experiments and in constant practice, yet after the lapse of even one year, when we have had occasion to repolish a speculum, there have been often disappointments. The difficulty arises from the necessity of employing two strata of resinous matter, one so hard, and both so thin. If in preparing the polisher the hard resinous composition is suffered anywhere to come in contact with the iron, the polisher will not retain its figure, and there will be a failure. A small chip of wood in the pitch will produce the same effect. If the water or lubricating mixture is supplied too sparingly, the polisher will begin to dry in spots, the rouge and abraded matter will collect there, and the thin stratum of pitch will be compressed till the accumulated matter resting upon the iron will act just as a chip of wood. If the lubricating fluid is a little in excess the rouge will run loose, the very hard resinous surface being able to retain but a very small quantity of it, and the incipient polish will disappear. An excess of rouge acts in the same way, while, if the rouge is not in sufficient quantity to keep up the cutting-action, the surface of the speculum loses its truth. The process therefore requires great attention throughout. Both the temperature of the water in which the speculum revolves and the temperature of the room, of course, must be properly regulated. The process does not proceed well unless the moisture between the speculum and polisher gently evaporates, so that drops of fresh fluid may be added from time to time, to carry away the undue collection of abraded matter. As the hygrometric state of the air varies, so will the quantity of fluid required to lubricate the surface; and that would be a source of considerable embarrassment, were it not that in dry states of the air the dew-point can be adjusted by a jet of steam. When the air is very damp we have no practical remedy; and therefore the operation is not then attempted.

We have often endeavoured to evade these difficulties by employing a surface less hard, supported by a thicker substratum of pitch; but there has been an evident sacrifice of ttruh of surface and figure, and we have failed in obtaining that very fine definition which resulted from the old process when perfectly successful. By the old process, a speculum of 3-feet aperture and 27-feet focus has been frequently made so perfect that in favourable states of the air it has defined sharply the dots and figures on a watch-dial distant 100 feet, the eye-glass being a single lens of one-eighth of an inch focus: such a speculum in ordinary weather perhaps does little more than one that is inferior to it, both, for instance, showing well the sixth star in the trapezium of Orion; but in extremely fine nights it displays its powers by resolving nebulæ in which no traces of resolution had been seen before, and by concentrating the light of minute stars and so rendering them visible.

If the vivid polish of a speculum employed in the *open air* was as enduring as that of glass, the difficulty of the process and its uncertainty without continual practice would have been no great objection to it; but when, on the contrary, it is necessary to repeat the process at intervals perhaps so long that minute details are not fresh in the memory, the task becomes the labour of Sisyphus.

A very fine speculum loses much of its light and some of its truth of surface by being repeatedly dewed, especially if it has been several times cleaned, and for the ordinary work we are engaged in will be inferior to a moderately good speculum which is quite fresh.

The preparation of a polisher in the way formerly described is one of the great difficulties; a certain degree of manual dexterity is required, which can only be obtained by practice and kept up by practice. For many years we have very often prepared it in an easier way: some pitch of the proper consistence for polishing at 55° is put into warm water; and when soft, a little is taken out and rolled upon a wet board to the proper thickness. There is no difficulty in this, as the roller is governed by ledges of proper height at each side of the board. The surface of the pitch is wiped dry, and a thin stratum of the hard resinous composition in powder is sifted on. A large flat-iron, red-hot, is then passed over at a distance of 3 or 4 inches, and so slowly as just to fuse the resinous powder without making any change in its composition. The pitch, so prepared, is cut into squares of the proper size, and thrown into cold water till required. The polisher, warmed to about 80°, is then brushed over with very soft pitch, and, when the temperature has fallen to about 65°, the square pieces of pitch are arranged in their places and soon become quite fast. The whole of this operation requires little experience, and can be managed by common workmen—a great advantage. It has, however, this disadvantage, that the pitch is somewhat thicker than we should wish. In the 6-feet polisher the squares are 2½ inches; and although the soft pitch in the circular grooves will no doubt yield a little, still we have a larger continuous surface than we had by the original process, and therefore the pitch requires to be thicker. The reason why with the long transverse strokes the pitch must be so thin is evidently this, that the polisher passes so far beyond the edge of the speculum. If we coat a polisher with pitch alone and of some thickness in the ordinary way, and then proceed to polish, we shall find that, if at any time we stop the machinery for a few moments when the polisher is at the extremity of the stroke, the pitch will change its figure. The change of course will be less as the stroke is shorter; but by prolonging the time, even with a very short stroke it will still be perceptible. So considerable is the change of figure under such circumstances, that after some time a distinct mark will be made by the edge of the speculum, and the projecting portions of the pitch becoming comparatively protuberant, unusual force will be required to effect the next stroke. These continual changes of figure, slight as they may be, will produce excessive action on the outer portions of the speculum. To meet this evil, if we diminish the length of the strokes much, we impair the self-correcting action to which we are mainly indebted for success. To explain this, let us suppose the throw of the first eccentric, B (see figure in 'Transactions' for 1840), to be reduced to a small quantity, and the action of the second eccentric, G, to be reduced in the same proportion, the speculum continuing to rotate; if the polisher and speculum are not truly portions of the same sphere, there will be unequal action at the centre or edge of the speculum, according as the polisher is more or less convex than it ought to be. In the first case, a depression will be formed at the centre of the speculum of a diameter proportional to the throw of the eccentrics; in the second case there will be an annular depression at the edge of the speculum. It is plain that the speculum cannot be restored to truth till the remainder of the surface has been lowered to the depth of the depression: this, however, will not be accomplished in practice if unequal action is continued even for a very short time. It may be thought that rigid identity of figure might be secured in the first instance; but this is practically impossible: the rouge cannot be distributed with perfect regularity; besides, as the temperature of the polisher varies, so does its radius of curvature. But even if perfect coincidence was secured at the beginning, it would not long continue. With very small motions the abraded matter would not be equally distributed, and, collecting in excess in some place, unequal action would be set up before the pitch had time to yield. If the excess was not at the centre, the depression would assume the character of a ring. The pitch at length yielding, the ring would not increase, but it would continue, and, a similar cause arising in another part of the polisher, a second ring would be formed, and so on. I have seen a surface of an annular character all over, the breadth of the rings depending on the adjustments of the eccentrics. Why the depressions once formed continue with so much persistency is evidently owing to the yielding character of the pitch, which, when the depression is of large area, becomes protuberant, precisely as it does where it overhangs the speculum, and so the cutting action is to some extent continued. An annular surface is produced by grinding, under similar circumstances, but the rings change their places frequently. The annular surface is always best-marked when the action of the second eccentric is suspended completely. To see the annular surface, the speculum must be slightly polished by rubbing it all over for a few minutes with a small lap covered with soft pitch and rouge. I need, perhaps, hardly add that the character of these surfaces can only be seen when they are examined by the light reflected from the watch-dial, in the way described in the 'Transactions' for 1840. As the throw of the eccentrics is increased the rings gradually disappear; and when they reach the proper positions the surface becomes quite uniform. We have often found it very useful, when the figure of the polisher was not satisfactory, to throw another movement into gear connected with the guide D, by which an occasional stroke was given of increased length: the cause of unequal action is thus immediately removed if it does not arise from some defect in the construction of the polisher, such as the contact of some unyielding substance with the iron. The experiments I have just referred to were of a very early date, but they were numerous and made with great care; I have therefore not thought it necessary to repeat them.

In the first polishing-machine we made, the polisher was connected with the eccentric B by means of a rigid bar passing through the guide D, the guide being furnished with an adjustment at right angles to the line joining the centres of the speculum and eccentric. The guide was equidistant from the centres of the eccentric and polisher, and the path of a point in the polisher was similar to that of the crank-pin of the eccentric. We found, however, that when the movements were very small the surfaces both in grinding and polishing became somewhat annular, and when the movements were large the figure was spoiled. We therefore substituted a jointed rod for the rigid bar, and added the second eccentric. From time to time we have returned to the rigid bar, tempted by its obvious advantages, and hoping in some degree to free it from its defects. It is an advantage that with it the movement of a point in the polisher is as the circumference of a circle, while with the jointed bar under similar circumstances it is as the diameter. In the one case there will be more than three times the amount of motion there will be in the other, and the polisher will overhang the speculum but for a moment at each stroke, instead of dwelling for a much longer time twice during each revolution of the second eccentric, and therefore there is not the same necessity for employing a very thin substratum of pitch; the process therefore is a much easier one. We have found that with the rigid bar, and, indeed, with the jointed one, a slight periodical movement of the guide D contributes much to free the surface from an annular character, for reasons which are obvious. The guide D is mounted now like the eccentric G, and a band from a small pulley on the axis of B, acting on a large one on the axis of D, effects the object in a very simple way.

When a speculum has been truly *ground* by the machinery acting with transverse strokes, the rigid bar will polish it on very easy terms, and for all the ordinary work of the observatory it will be sufficiently perfect.

We have long been in the habit of resorting to the rigid bar when out of practice and we required at once a fresh speculum.

A speculum of 3-feet aperture, which has usually been uncovered in all weathers for visitors, has frequently been so polished, and it has borne well a quarter-of-an-inch lens when tested with a watch-dial while on the engine. When a speculum of 6-feet aperture was last polished a rigid bar was employed; and the result was tolerably successful.

Since the publication of Mr. LASSEL's experiments we have several times tried simple pitch, the movement being given by the rigid bar, but we have not succeeded in obtaining as good a surface or as fine definition as when the polisher was prepared in a more elaborate manner.

The combination of soap and ammonia which we employ may be prepared in this way. Half a pound of brown soap is dissolved in one quart of warm water, and one quart of strong water of ammonia is added. The bottle is then corked and shaken from time to time, for a week at least: we think it improves by keeping. One ounce of this mixed with eight ounces of water makes the lubricating fluid. The mixture should be made the day before, and kept in an open vessel, so that the excess of ammonia may evaporate. We were at first apprehensive that in employing this mixture we were endangering the hard film of the polisher, and so perhaps sacrificing to some extent truth of surface; but this was found not to be the case unless the ammonia was much in excess. As a kind of experimentum crucis, we polished specula with simple pitch rather softer than usual, employing pure water and the saponaceous mixture alternately, and found that the mixture was favourable to truth of surface instead of the reverse.

We have long ceased to make rouge, as it can be obtained of good quality from the rouge-maker.

In shaping the polisher by applying it to the speculum, we find it better that the polisher should be quite cold, while the pitch and resinous composition are slightly warm. We pass the flame of a few shavings or of wood-spirit under the polisher with its surface down, and instantly apply it to the speculum for half a minute. This is repeated till there is satisfactory contact. When the polisher was warm, we found it was difficult to avoid compressing the pitch too much. A crane makes the 3-feet polisher quite manageable; and a travelling crane, with railway overhead and screw, effects the same thing for the 6-feet polisher. Both polishers are provided with gimbals, so that they can be instantly turned over. Though it is better to prepare the polisher fresh each time, we have often employed the same polisher successfully two or three times. In that case the polisher must be washed, and when dry the surface is to be very slightly moistened with spirits of turpentine. A thin film saturated with rouge will thus be removed; and a flame passed under will evaporate the turpentine. The polisher is then to be inverted and warmed to about 80°, the face being uppermost and again turned over. If now a flame is employed cautiously two or three times at intervals, the pitch at each square will become protuberant, bearing the hard resinous film on its surface, and the polisher will be restored very nearly to the same state it was in when originally prepared.

No one will be so ill advised as to attempt to construct a large reflecting telescope without first collecting all the information to be obtained in books. In Mr. LASSEL's paper in the 'Transactions' of the Astronomical Society he will have an excellent guide. Should he employ an apparatus similar to ours, the speculum is first to be truly ground with the jointed bar. The throw of the first eccentric is to be one-third the diameter of the speculum, and that of the second, measured at the speculum, about one-fourth. It will be better in all early experiments to rely on pitch alone, carefully adjusted to the temperature at which it is to be used, perhaps 55°. The jointed bar which was employed in grinding the speculum is to be exchanged for the rigid bar, the eccentric and guide being readjusted. When the speculum has been successfully polished a few times in this way, an attempt may be made to obtain a better result by facing the polisher with a hard resinous composition; and finally the jointed bar may be resorted to, but at the same time the thickness of the pitch must be greatly reduced.

As to the mounting, it is simple, and any engineer could execute it without difficulty, the photographs supplying the necessary information.

The tube is supported at its lower extremity by a massive universal joint. It is counterpoised by weights which are constrained to move in a circular arc which nearly coincides with the curve of equilibrium; and a steady strain is kept upon the suspending-chain by means of three weights attached to levers, which successively come into play as the tube approaches the zenith and passes north beyond it: the levers are about two-thirds of the length of the tube, and have cross heads at their lower extremities, which are formed into bearings; and when in their places, the cross heads are all parallel to each other and to the transverse axis of the universal joint, from which they are about 5 feet distant. The levers thus move steadily in one plane, that of the meridian. A chain connects the levers at the proper intervals and the tube with them; and as the tube descends, each lever takes its place successively in a deep recess in the ground, the chain subsiding into a heap. This contrivance is effectual, and the chain has never fouled. The three weights are of different sizes, so proportioned as to reduce as much as possible the deviations from exact equilibrium at different altitudes, due to the irregular action of the counterpoises, which move in an arc of a circle instead of the proper curve. A slow hand motion was originally fixed near the mouth of the tube, for raising or lowering it in taking measures; but we do not find it necessary. The telescope at the equator can conveniently follow an object three-quarters of an hour; and its motion is nearly equatorial: it would be almost exactly so if the pulley of the suspending-chain was in a line drawn from the axis of the universal joint parallel with the axis of the earth. The pulley, to give the chain more mechanical advantage in raising the telescope when very low, was placed above this line; but there was at the same time an arrangement made so that the chain might be brought, when necessary, by means of a grinding pulley, into the proper line. In practice we have found the movement of the telescope sufficiently equatorial without this: at a little distance, however, from the meridian, the plane of the position-circle of the micrometer deviates sensibly from a plane passing through the pole in all positions except when near the equator, as will be evident on considering the construction of the universal joint; and the distance from the meridian must be known where much precision is required in the reduction of the observations.

The motion in right ascension is by a rack the extremity of which bears by rollers on a circular arc of 40-feet radius. This rack is connected with the tube by a pinion, and the pinion is acted upon by an endless screw driven by a pulley, which pulley is driven by a band from a porter's wheel attached to the lower end of the tube. The pulley can also be moved by the observer; but this is not often necessary. The large circular arc is in pieces 5 feet long each, carefully planed, but not touching at their extremities, to guard against unequal action. The surface of each segment was adjusted separately to the plane of the meridian by a transit-instrument. And thus the means were provided for taking right ascensions with considerable accuracy. For polar distances there is a circle, 18-inches radius, at the lower end of the tube, furnished with a spirit-level; but for finding objects, there is an index of 6-feet radius connected with the transverse axis of the universal joint; so that the instrument can with the utmost facility be set roughly in polar distance. Means, of course, are provided for enabling the observer to reach the eyepiece in every position of the telescope. From 120° of polar distance to 80° this is effected by a stage, nearly counterpoised, which slides on bearers, the observer standing in a small gallery, to which he can communicate a transverse motion: the large stage is raised by a windlass; and, to guard against possible accident, there is an arrangement which locks it, completely under the command of the observer. From 80° to 50° the eyepiece is reached from the second gallery, and from 50° to 25° by the third. The fourth gallery, reaching to the pole, for which machinery was made at the same time as for the others, has never been put up, the other galleries so far furnishing ample employment for the instrument. The action of the galleries, and the way they are secured, is sufficiently evident from the photograph: they are of great strength, and in their construction, as well as in all other parts of the instrument from which there seemed to be a possibility of accident, the ordinary engineering rules as to strength have been considerably exceeded. The eyepiece arrangement consists of two adapters fixed into one slide, so that there are always two eyepieces, a high and a low power, which may be employed successively simply by moving the slide. The slide is counterpoised, and the eyepieces fit in without screws. The telescope is perfectly steady even in a high wind, and we have had occasionally very fine definition during a strong gale.

From the experience we have now had, I think I may safely say that, where objects are to be observed only at short distances from the meridian, this plan of mounting is convenient and effective, and I do not see room for any material improvement. Where observations are carried on systematically, I do not think there is very much disadvantage in the limited movement in right ascension. Objects are best seen near the meridian, and no object can be thoroughly examined in any other position. There is, indeed, a small portion of the heavens which can scarcely be observed in the perfect absence of twilight, and an object there situated would probably be better seen, even at some distance from the meridian, when it was perfectly dark; but that seems to be of little moment. The really important disadvantage of the limited equatorial movement seems to me to be this, that, where fine nights are extremely rare, with an instrument so limited it is impossible to turn them to the best account.

It now only remains to answer in the best way I can the questions—

First, What magnifying power can be usefully employed with a speculum of 6-feet aperture? I perceive, in looking through the observations, that the single-lens eyepiece, ½-inch focus, being a power of about 1300, is often mentioned as giving better vision than lower powers. That, I presume, may be considered the highest power it has been found advantageous to employ in general observations upon the nebulæ. With the speculum of 3-feet aperture I have occasionally employed powers exceeding 2000 in bringing out minute stars; and the speculum of 6-feet aperture has sometimes been in sufficiently good order to admit of equal or perhaps higher powers; but in our climate the opportunities of employing such powers are rare, and of short duration.

Secondly, it has been asked, Could a telescope be made of larger dimensions, and would it be of service? I feel little doubt that both questions should be answered in the affirmative. A speculum of larger aperture would, probably, on favourable nights bring out faint details of interest in the nebulæ, and add to the number of known double and multiple nebulæ. Something, however, will perhaps be accomplished in that direction in our future observations, by employing silver for the second reflexion; but if ever telescopes of equal power are erected in climates more favourable than this, perhaps more will be effected than would be possible here by pushing increase of aperture to the largest practicable limits.

In making a selection from observations so numerous, there has been considerable difficulty. It was not always easy to decide how much it would be practicable to omit without the danger of conveying an erroneous impression—without, on the one hand, perhaps unduly weakening the evidence of the fact recorded, or, on the other, unduly strengthening it. For instance: if, in the observations of a particular object, we find it recorded on six different nights that a minute star was seen involved, and that on six other nights it was not seen at all, the twelve observations extending over a series of years, two cases might arise requiring very different treatment. First, if the nights when the star was visible were irregularly interspersed between the nights when the star was invisible, and we saw enough in the state of the atmosphere or speculum to account for the occasional invisibility of the star, it would probably be quite sufficient to enter one good observation when the star was distinctly seen. Suppose, however, that the nights when the star was seen were all included in the observations of the first three years, but that the nights when the star was not seen were in the observations of the last three years, then it would be necessary to enter all the observations, so that each person might be enabled to form his own opinion as to the cause of the discrepancy. 838 H., fig. 11, in the 'Transactions' for 1850, is a remarkable instance of this: from 1850 to 1858 the small star was not seen.

The details of faint nebulæ with curved or spiral branches have usually been made out by degrees, not only on successive nights, but often in successive years. In such cases we have not usually thought it necessary to give the early observations, or the observations on unfavourable nights, but merely a few good observations embodying the whole amount of information we had been enabled to obtain.

New nebulæ have not been looked for, our object being to scrutinize the more promising of the old ones; but new nebulæ have often been found in their immediate neighbourhood, and their places have usually been entered roughly in the observing-book, and a slight diagram made in the margin, so as to ensure their being easily found again: in such cases we have, to save space and diagrams, merely written "novæ near," and have only entered observations when the micrometer was employed. We have also, for the same reason, omitted many diagrams of known objects, where the positions and distances were merely estimated.

In the case of each object, we say "observed so many times;" that means that we have recorded observations of it on so many different nights: it may have been seen frequently on other occasions.

Where an object has been marked "observed several times," and nothing more, the inference is that with an instrument such as ours is, and in our climate, it would be waste of time to examine it further in the hope of making out details of interest.

It will be observed that the cases are very numerous where stars have been seen on the edges of nebulæ: we have taken care to enter each case, often, however, on the authority of a single good observation, as before explained.

The words "mottled" and "patchy" mean the same thing. Where the nebula is of that character, it is worth examining under favourable circumstances. The faint spirals have often been first seen as "mottled."

The word "finder" means the eyepiece with a large field. The telescope has no finder in the common acceptation of the word.

The letter "r" has been occasionally added to the description, and always in the same sense as that in which HERSCHEL employed it: I do not, however, attach much importance to the expression of opinion it conveys, because the question of resolvability can only be successfully investigated when the air is steady and the speculum in fine order. In the early observations with the 6-feet telescope we had the advantage of a very fine speculum; it had been polished at the close of a long series of experiments with 3-feet specula, when by practice every refinement of manipulation was fresh in the recollection; there were also at that time several very good nights; and many nebulæ were resolved. Very soon after, the spiral form of arrangement was detected; and our attention was then directed to the form of nebulæ, the question of resolvability being a secondary object. In the mean time the speculum, which had been frequently dewed and occasionally cleaned, had lost its fine edge, and was no longer in a state to deal with the question of resolvability. Our aim was to trace out faint details, and in that respect also the speculum had lost much of its power It was therefore repolished, and, though less perfect than before, did the work we required well. Since that, we have had perhaps two or three specula as perfect as the first one; but the mass of observations have been made with specula considerably inferior to it, and, I am sorry to say, very often not as bright as they should have been. The removing a 6-feet speculum from the telescope to the laboratory, repolishing it, perhaps several times, and replacing it, is a serious operation, and has often been too long postponed. While the telescope was in constant use in all weathers, it would have been a hopeless task to attempt to keep it in a state fit for the resolution of nebulæ, and the attempt was not made. I may, perhaps, mention that with the 3-feet speculum in fine order I have often detected resolvability when there was no trace of it with the 6-feet speculum in its ordinary working-condition.

The question of resolvability, therefore, I think, must remain to be taken up separately, when the finest instrumental means are available, and when it may no longer be necessary to subject the specula to the wear and tear of constant work.

As to the nebulæ which have nuclei, some are described as increasing in brightness very gradually to the centre, others very rapidly, and some as having a stellar nucleus, or perhaps a star in the centre. These descriptions, however accurately conveying the impressions made upon the eye at the time, cannot be taken as in all cases representing real physical facts. A star may have been mistaken for a condensed nucleus, or the reverse; and it is often impossible to say which of the two suppositions is the more probable. The remarks as to the question of resolvability apply with equal force to the questions relating to the structure of nuclei. It is, however, probably worth remarking, that, while amongst the clusters there are objects which, if removed to a sufficient distance, or examined with an instrument of insufficient power, may be supposed to be representations of nebulæ with centres of varying brightness or condensation, there seems to be no cluster with a central star of such surpassing magnitude that under any circumstances it could be taken as the representative of the class of objects described as having a star in the centre.

The little rough sketches in the margin are exact copies from the sketches made at the moment in the observing-book. There are usually several sketches of the same object made at different times; we have endeavoured to select the best.

As to the drawings, they usually represent the objects a little stronger than they appear on an ordinary night, but not stronger than on a fine night, when the air is clear and the sky black. Most of them have been repeatedly compared with the objects by different persons, and some have been several times sketched independently; so that I trust they are upon the whole accurate. The central portion of the nebula of Orion has been drawn with great care by Mr. BINDON STONEY, and Mr. HUNTER has been engaged this season in finishing the remainder; but another season will be required to complete the work.

Although there is probably no remarkable object in the list which I have not several times examined myself, for the great bulk of the observations I am indebted to the gentlemen at the time in charge of the Observatory. Mr. JOHNSTONE STONEY's observations commenced in July 1848; and in June 1850 he was succeeded by his brother, Mr. BINDON STONEY. He continued in charge of the Observatory till May 1852; after that, Mr. MITCHELL observed for about two years.

Mr. JOHNSTONE STONEY occasionally also worked with his brother, and sometimes with Mr. MITCHELL.

Though so many of the observations were made in my absence, they are not the less to be relied on: nothing was done by an unpractised hand, and no pains were spared to ensure accuracy. I refer with as much confidence to the observations of the two Mr. STONEYS and Mr. MITCHELL as if I had on every occasion been present myself, because I know that they had thoroughly mastered the instrument and the methods of observing before they recorded a single independent observation; they were, besides, eminently cautious and painstaking.

There are no micrometer observations by Mr. MITCHELL: I now rather regret it, as several cases of suspected change have recently been brought to light in arranging the materials of this paper. The fault, however, was mine. It appeared to me so highly improbable that any change would be detected, that I requested Mr. MITCHELL to press on and not spend time on the micrometer. The most remarkable case of suspected change is perhaps H. 1905. HERSCHEL gives a drawing of it, the axes of the two nebulæ in a line. On April 11, 1850, Mr. JOHNSTON STONEY remarks the two nebulæ not in a line. April 17, 1855, Mr. MITCHELL remarks the two nebulæ are not in a line, but the axes are parallel, and gives a diagram. At the present time they are neither in a line nor parallel, but inclined at an angle of about sixteen degrees. The micrometer is employed without illumination; various contrivances were tried for illuminating the lines in a dark field; but the darkness was not absolute, and faint details were obliterated. We therefore substituted bars for lines.

In the 'Transactions' for 1850 are given Mr. JOHNSTON STONEY's measures of H. 1622. M. OTTO STRUVE was good enough to send me measurements of the same spiral, and to direct my attention to the fact that Mr. STONEY's positions are about two degrees in excess.   A little consideration made it evident that the construction of the universal joint which supported the telescope was the cause of the error, and that a certain correction must be applied, depending upon the polar distance and the distance from the meridian.   Mr. BINDON STONEY's measures of H. 1622 are also given, as also his measures of H. 2060, and STRUVE's measures of both.   As OTTO STRUVE's measures are no doubt as exact as possible, it will be easy to judge of the degree of dependence to be placed upon the other measures made by us.

As to the figured nebulæ, little can be added to the information contained in the general catalogue.   Plates XXV. and XXIX. figures 7 and 35, are from sketches by Mr. HUNTER, but they had been previously sketched several times by others.   We preferred Mr. HUNTER's sketches, thinking they were upon the whole the most accurate, containing some additional details.   Figure 43, the Dumbell, by Mr. BINDON STONEY, is based on micrometrical measurements, and is thoroughly to be relied upon.   No stars are inserted which have not been measured.   The powers used were low, the ordinary working-eyepiece: with high powers the faint details vanish.

The original observations are in books, in which they were entered each night: from time to time they were copied into a folio in the order of right ascension; and of that folio a copy was made for ordinary use in the Observatory.   It will be easy therefore to supply, to any person who may be engaged in observing a particular object, all the information we possess.   We have not given the places of the objects brought up to the present day, but merely HERSCHEL's numbers, to save space.

It is hoped that further inquiry will be suggested by the questions raised in the following observations; they have already opened up to us new grounds for further research.

INDEX TO THE FIGURED NEBULÆ.—PLATES XXV. to XXXI.

Figure.	Number in the observations.	By whom drawn.	Figure.	Number in the observations.	By whom drawn.
1	15	Mr. Mitchell.	23	1306 & 1308	Mr. Mitchell.
2	156	Mr. B. Stoney.	24	1337	,,
3	232	Mr. Mitchell.	25	1385 & 1392	,,
4	241	,,	26	1414 & 1415	Mr. B. Stoney.
5	242	Mr. B. Stoney.	27	1441	Mr. Mitchell.
6	262	,,	28	1589	,,
7	311	Mr. S. Hunter.	29	1650	,,
8	315	Mr. Mitchell.	30	1713	,,
9	327	,,	31	1905 & 1906	Mr. S. Hunter.
10	131	,,	32	1946	,,
11	393	Mr. B. Stoney.	33	1968	Mr. B. Stoney.
12	421	,,	34	2075	Mr. Mitchell.
13	689	Mr. Mitchell.	35	1744	Mr. S. Hunter.
14	692 & 693	,,	36	2084	Mr. B. Stoney.
15	765 & 766	,,	37	2099	Mr. Mitchell.
16	875	Mr. G. J. Stoney.	38	2139	,,
17	1011	Mr. Mitchell.	39	2172	,,
18	1052 & 1053	Mr. B. Stoney.	40	2241	Mr. B. Stoney.
19	1061	,,	41	2245	Mr. Mitchell.
20	1111 & 1113	,,	42	2297	,,
21	1202	,,	43	2060	Mr. B. Stoney.
22	1245	Mr. Mitchell.			

EXPLANATION OF PLATE XXIV.

Fig. 1. The speculum upon its supporting levers, the apparatus by which the levers themselves are counterpoised having been previously removed to prevent confusion. 1. The lime-boxes, connected with the cover by sliding tubes. 2. The cover, which fits nearly air-tight. Before the cover is taken off, the lime-boxes are removed, all communication between the lime-boxes and speculum having been first intercepted by valves. Without this precaution, lime dust would make its way to the speculum. 3. Ring in which the speculum is suspended. 4. Supporting levers, which are shown in plan in fig. 10, where A represents a primary triangle, B a primary and three secondary triangles, and C one-third of the system complete, the dots being the balls. 5. The frame upon which the levers rest: a single casting. 6. A massive casting, which is bolted to the bottom of the table as soon as the speculum is in its place: it bears the wrought-iron girder (7) which holds the suspending-ring. 8. Turn-table for changing the specula: it is at the north of the telescope, close to the elevating-windlass.

Fig. 2. The same, with the weights and levers for counterpoising the secondary triangles. 9. The weight. 10. The common fulcrum of the levers. The levers are connected with the secondary triangles by slight straight rods.

Fig. 3. Grinding- and polishing-tool, seen at the back. It is suspended by gimbals; the tool can therefore be turned over easily from time to time, which is necessary in applying the pitch and resinous composition.

As soon as the tool is prepared the gimbals are removed; and it is then managed by the shackle in the centre. The shackle carries a triangle, with a lever at each angle. Each lever carries similarly two levers, connected at their extremities with T-shaped levers, which carry the tool by thirty-six points. The chain through which the counterpoise acts during the progress of grinding and polishing is hooked to the shackle, and the strain is thus distributed so equally that there is no sensible distortion.

Fig. 4. The same tool, seen in front. The straight grooves are produced by casting; and in this there is no difficulty, provided the pattern is nicely made. The little square blocks are kept in their places in the usual way by pins, and, remaining in the sand, are removed separately. The circular grooves are of course cut in the lathe.

Fig. 5 represents the telescope seen from the south-east, the stage of the first gallery being slightly raised.

Fig. 6, from the south-west, showing the machinery of the second and third galleries.

Fig. 7, from the south, showing the position of the telescope when a man enters the tube to fix the small speculum, and remove the cover of the large one, in preparing for the night's work.

Fig. 8, as seen from the north.

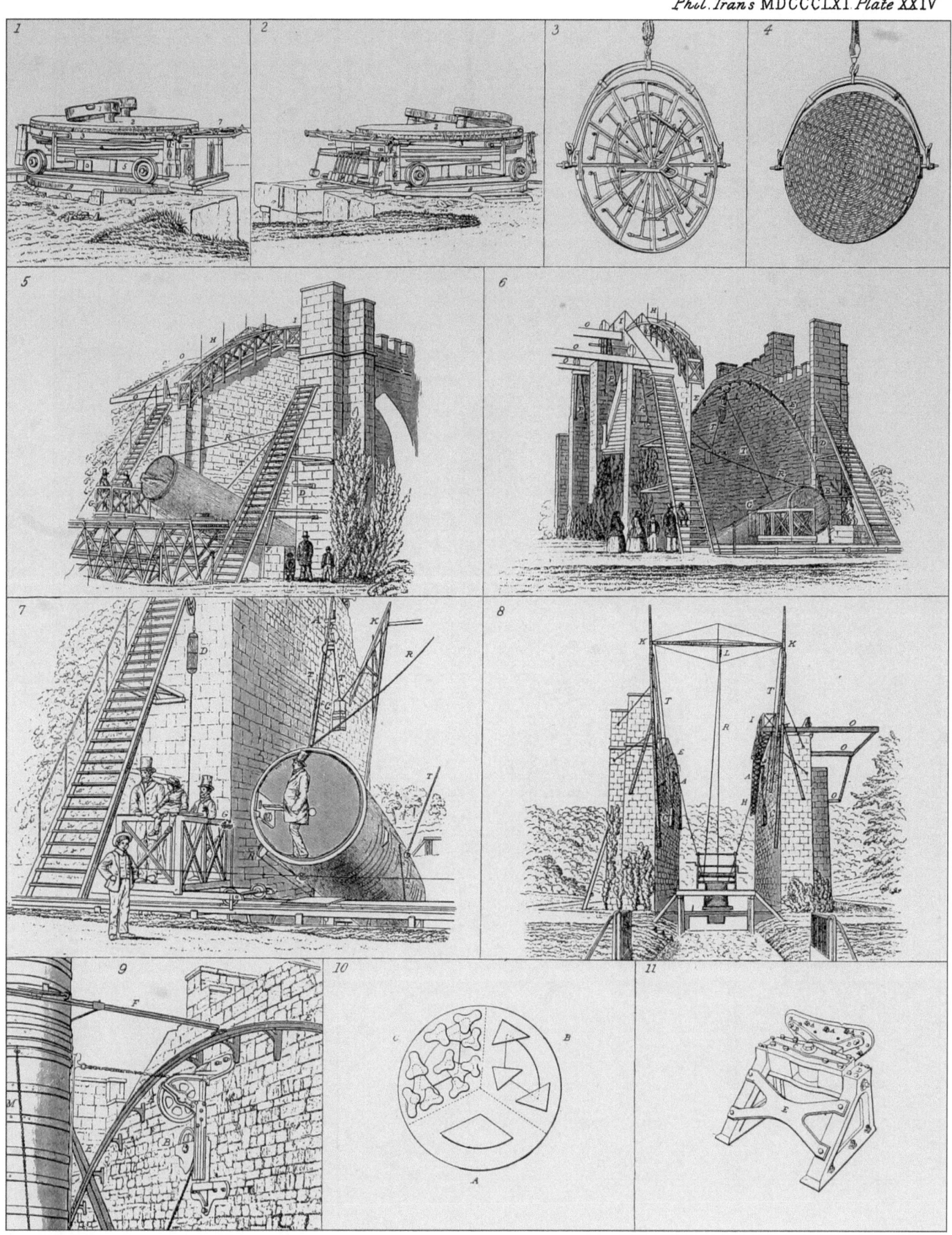

J. Basire. sc.,

S

N

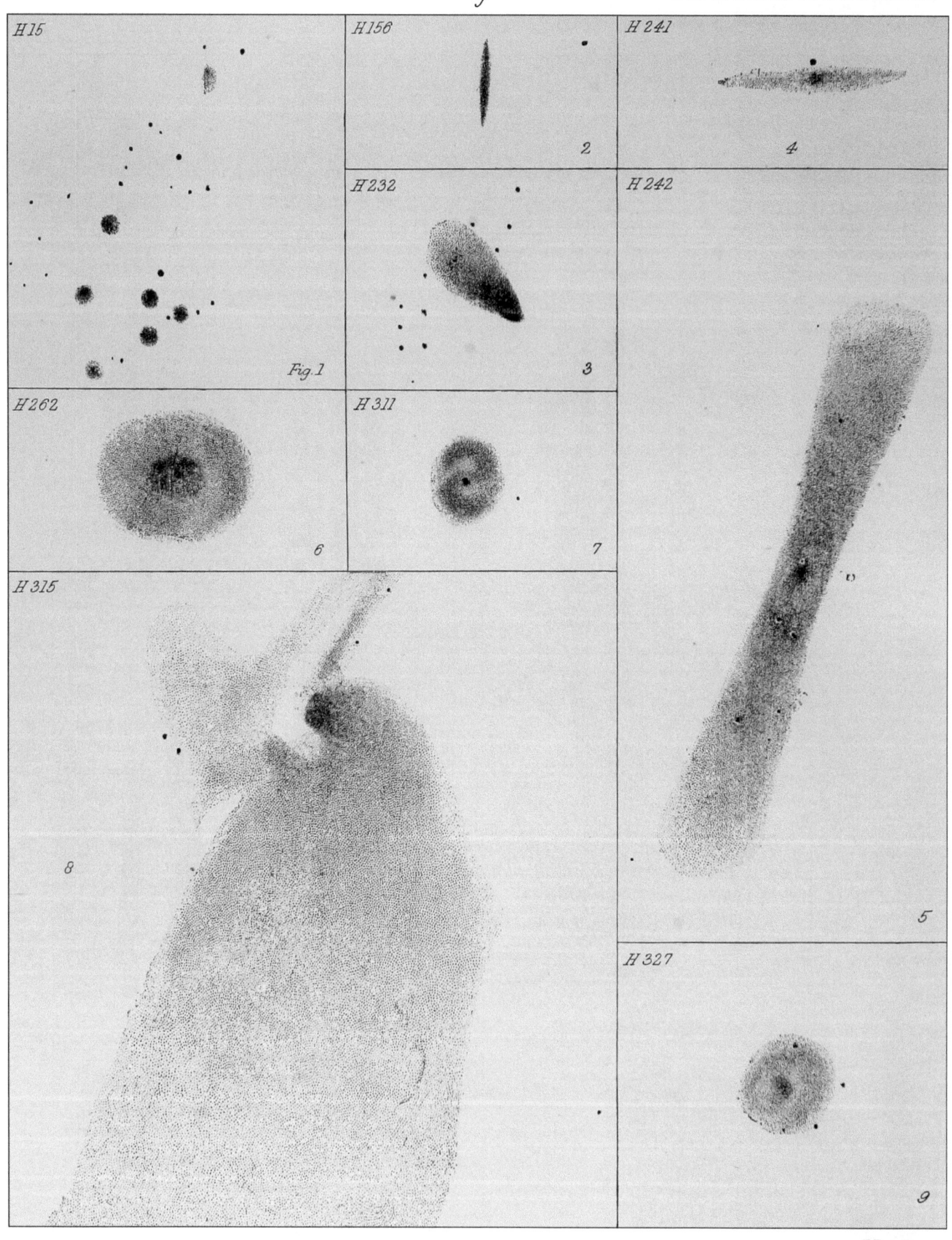

J. Basire sc.

H 131

Fig 10

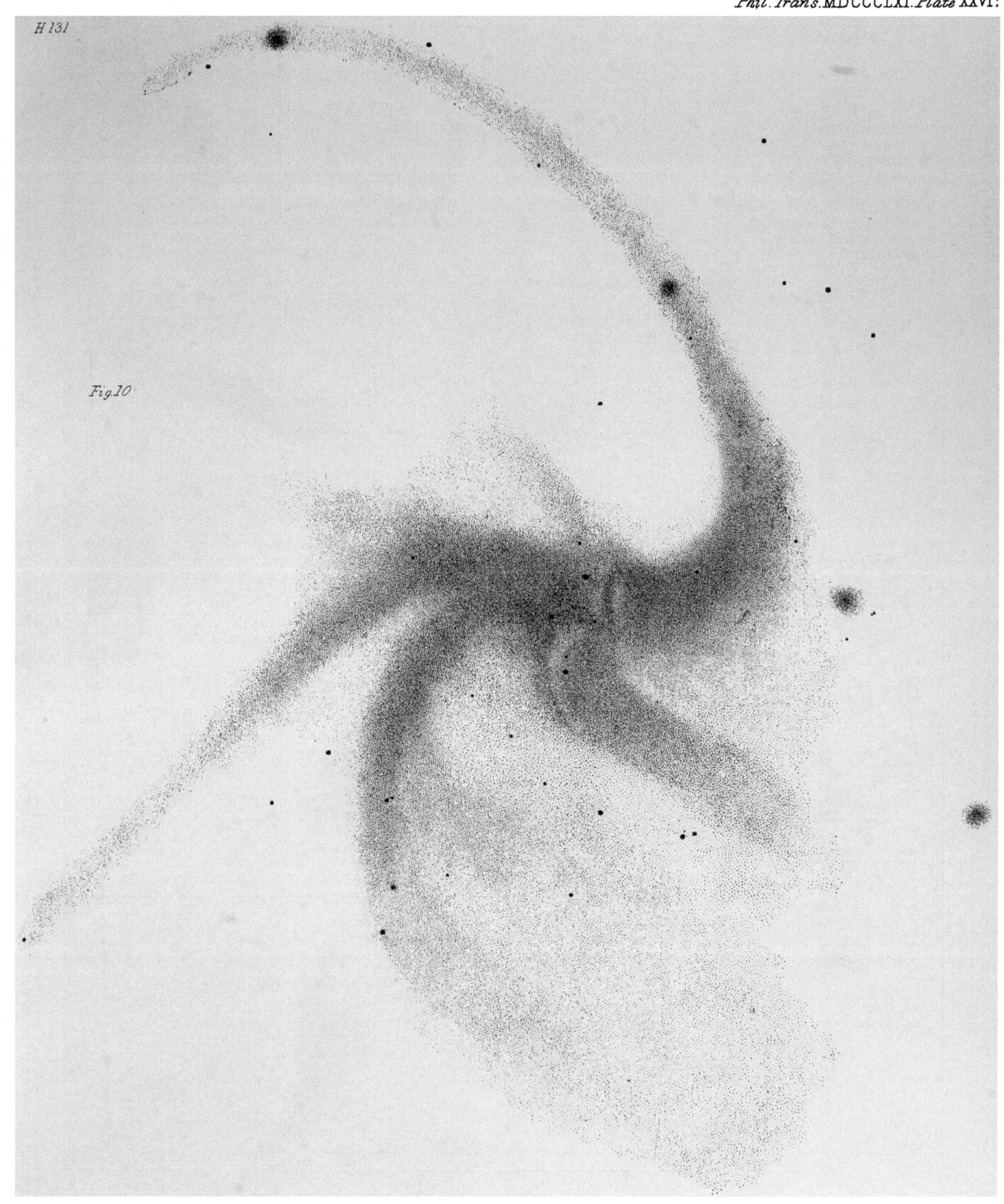

3

J. Basire. sc.

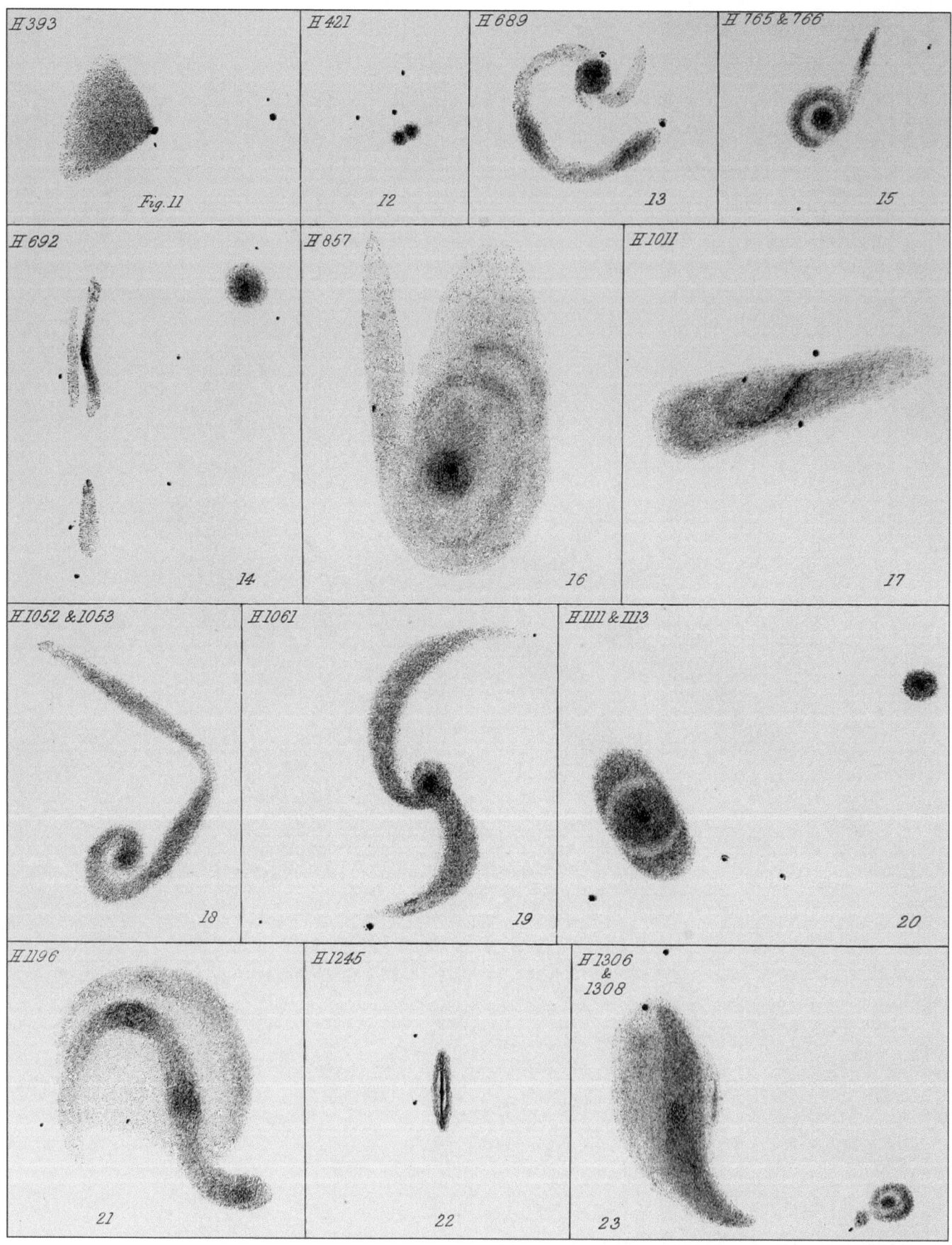

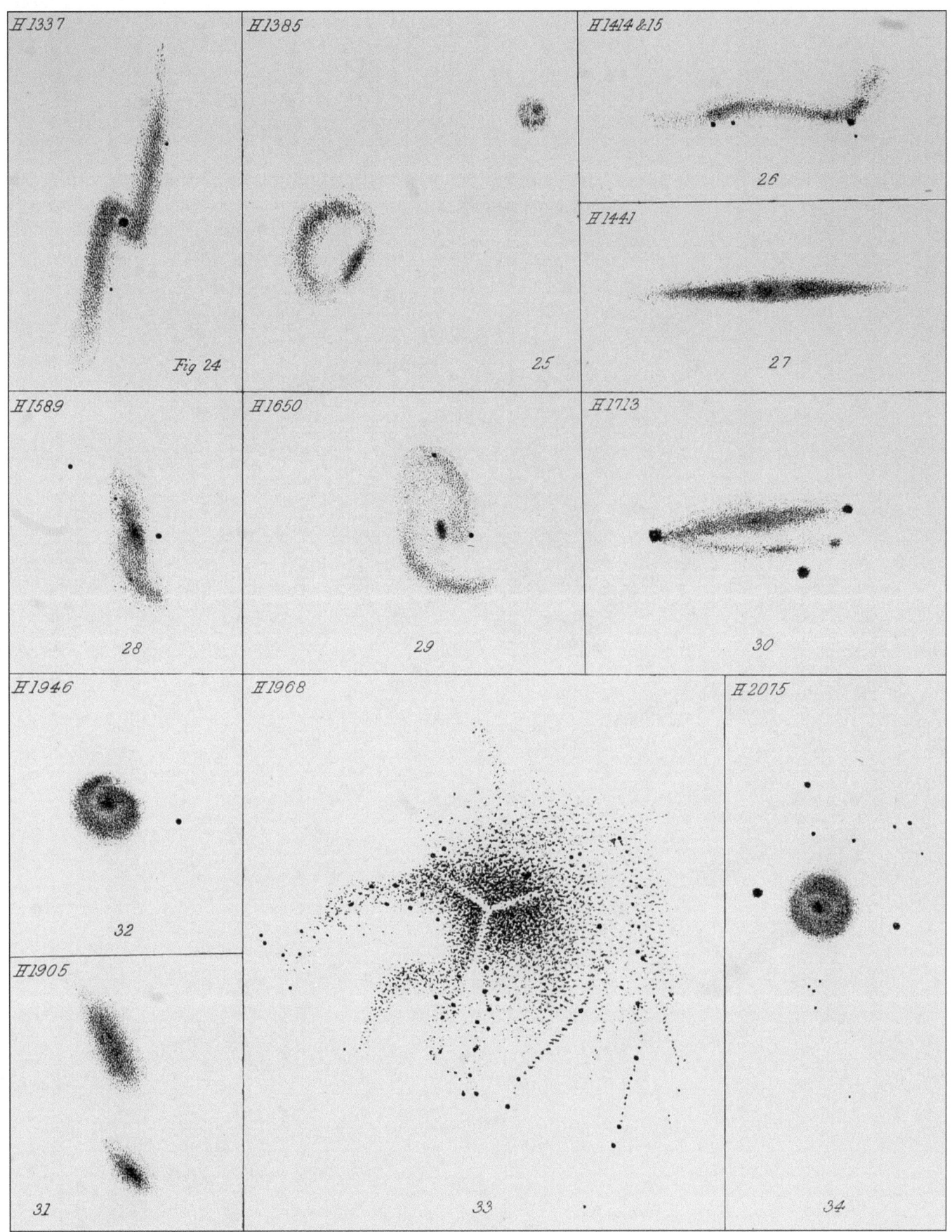

H 1337   Fig 24

H 1385   25

H 1414 & 15   26

H 1441   27

H 1589   28

H 1650   29

H 1713   30

H 1946   32

H 1905   31

H 1968   33

H 2075   34

J. Basire sc

Phil.Trans MDCCCLXI.Plate XXIX

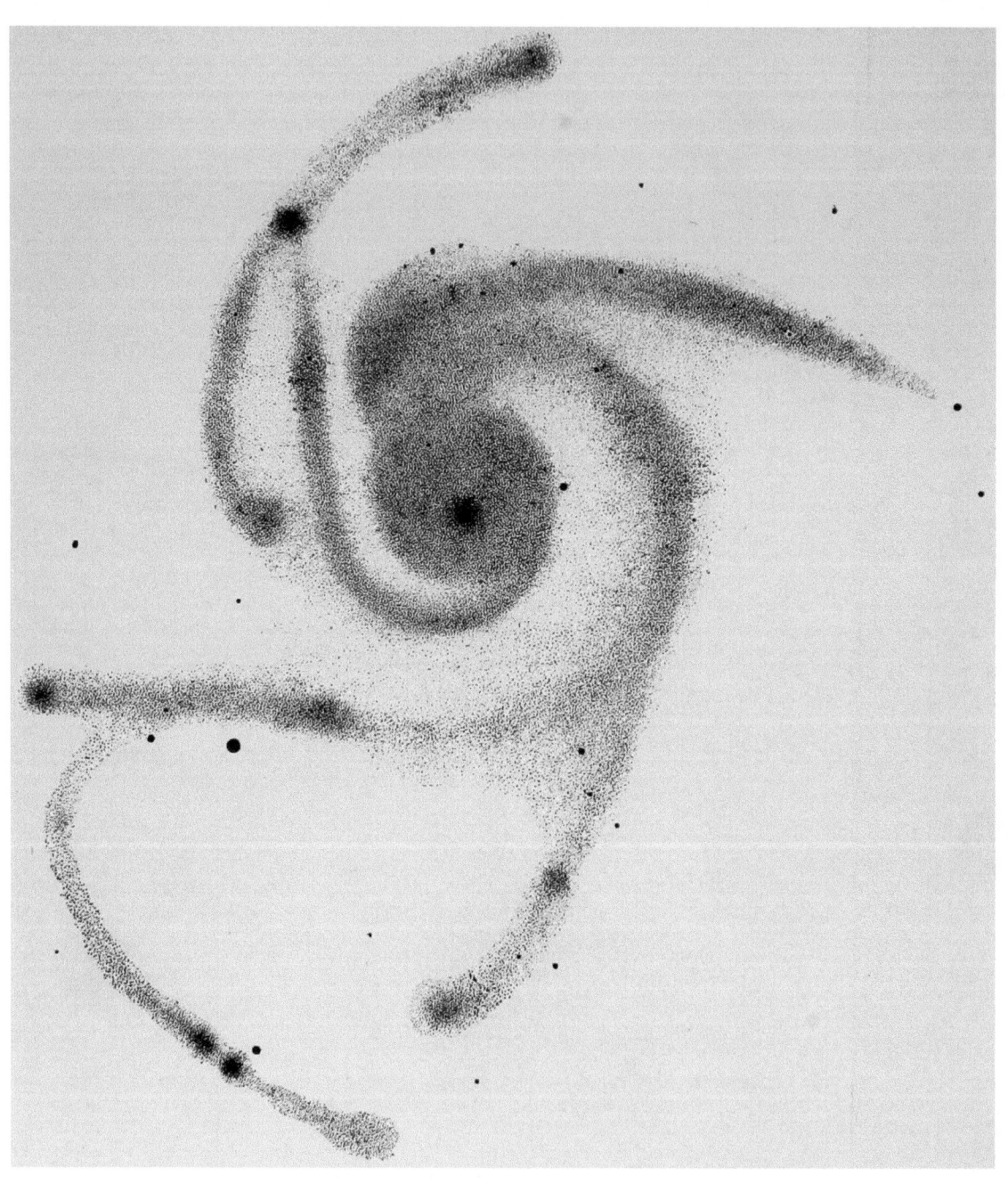

Fig.35.

J.Basire.sc.

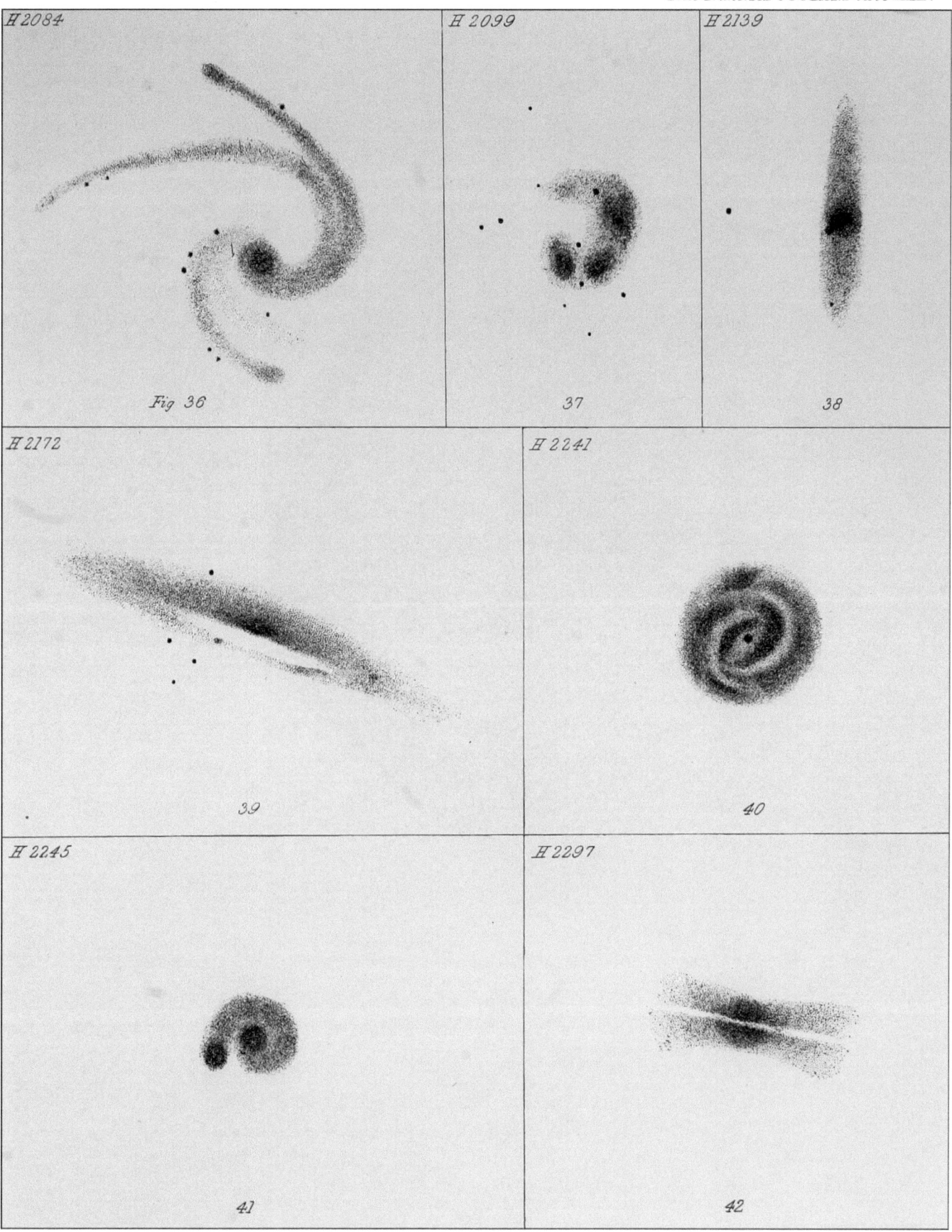

H 2084

Fig 36

H 2099

37

H 2139

38

H 2172

39

H 2241

40

H 2245

41

H 2297

42

J.Basire sc

Phil. Trans. MDCCCLXI. Plate XXXI

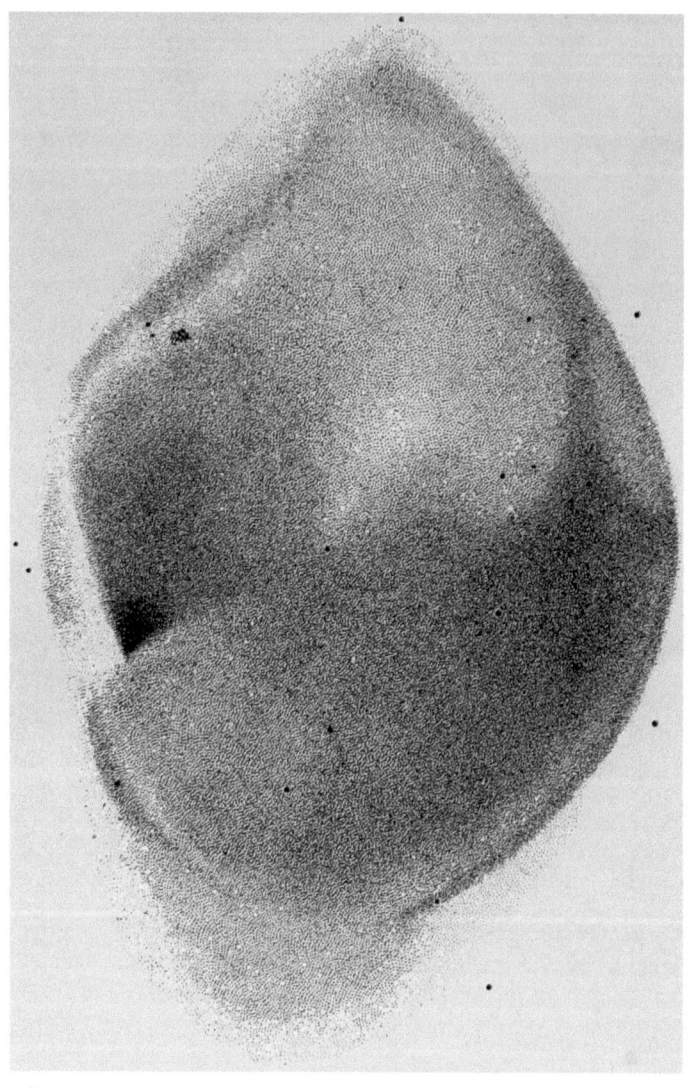

Fig. 43

J. Basire sc.

Fig. 9. The crane which carries the eastern counterpoise, on a larger scale.

The same parts are similarly lettered in all the figures. A, the cranes which carry the counterpoises. B, fig. 9, a guide-wheel by which the chain is kept in the axis of the crane in all positions of the counterpoise. The points of the shaft are placed eccentrically, so that it clears the wheel B when the telescope passes the zenith and moves north. F is of wood plated with iron, and is connected by rollers with the arc E, and by a rack and pinion with the tube. The pinion is driven by a wheel acted upon by an endless screw on the bar which carries the pulley N, fig. 7. The pulley N is driven by a band, and porter's wheel on the lower end of the tube, and thus the motion in right ascension is given. C, the principal counterpoises. D, the counterpoises of the stage of the first gallery. The stage is raised by increasing the action of the counterpoises D D. This is effected by chains attached to them, which pass underground to a small windlass. The counterpoises being rather less than the weight of the stage, it descends when the chains are relaxed. The stage carries the gallery G, which traverses on it in right ascension, motion being given to two of its wheels by a winch in the hand of the observer. E, an arc of cast iron made in pieces 5 feet long, not quite touching at the extremities, to guard against unequal expansion. Each piece was planed and accurately adjusted in the meridian by a transit-instrument. H, the second gallery. I, the third gallery. The galleries are supported by the beams o, which are plated with iron, and they are moved in right ascension by rack and pinion. P, tension bars to secure the iron framework carrying the wheels against which the beams press as the galleries are moved out. K, the cranes which carry the guy-chains, by which the counterpoises are constrained to move nearly in the curve of equilibrium. R, the chain which, passing over the pulley L, moves the telescope in polar distance. T, the chain of the principal counterpoise. In fig. 9 the chain M is seen, which raises three levers, each carrying a counterpoise. These levers successively coming into play as the telescope approaches and passes the zenith, maintain the chain R in such a state of tension, that the telescope obeys the windlass in every position. It has been found practicable so to adjust the levers that the residual error of compensation arising from the imperfect action of the guy-chains has been rendered almost insensible.

Fig. 11. Universal joint which bears the tube. A, bolts which secure it to the bottom of the tube. B B, two of the three adjusting-screws: these act against the carriage (5, fig. 1), directly under the ball-and-socket joint of the primary triangles. C, the axis perpendicular, and D the axis parallel to the horizon. The axis D gives motion to a large index for roughly setting the telescope in polar distance. E, a strong trussed framework of cast iron, secured to a solid stone foundation.

L

Abbreviations used in the Description of the Nebulæ.

B   denotes bright . . . .   thus pB means pretty bright.

b   ——— bright, or brighter . . .   —   pbM means pretty bright in the middle. (Large B applies to the neb., small b to a part.)

br   ——— broad . . . . —   40″ br = forty seconds broad.

cl.   ——— cluster.

d   ——— double   . . . . —   d✱ = double star.

E   ——— elongated . . . . —   E. n. and s = elongated in the direction north and south.

e   ——— extremely . . . . —   eF = extremely faint.

F   ——— faint.

f   ——— following . . . .   . . —   the f. of two = the following of two.

fig.   ——— figure.

g   ——— gradually . . . . —   gbM = gradually bright in the middle.

L   ——— large   . . . . —   pL = pretty large.

l   ——— little . . . . —   lbM = little brighter in the middle.

M   ——— middle.

m   ——— moderately . . . . —   mbM = moderately bright in the middle.

neb.   ——— nebula.

neby.   ——— nebulosity.

n   ——— north.

p   ——— preceding (when by itself or combined with n or s) . . . . —   np = north preceding.

p   ——— pretty (in other cases) .   —   pB = pretty bright.

pos.   ——— angle of position.

R   ——— round.

r   ——— resolvable.

S   ——— small.

s   ——— south (when alone or combined with p or f); suddenly (in other cases)—   sf = south following ; sbM = suddenly brighter in the middle.

v   ——— very . . . . —   vF = very faint.

vv   ——— extremely . . . . —   vvF extremely faint.

SELECTED OBSERVATIONS OF NEBULÆ.

Number in Herschel's Catalogue.	Number of times observed.	Description.
1	2	Nov. 4, 1850. Some stars seen in it, it is vF. Nothing further remarkable.
2	1	Sept. 18, 1857. 2 neb. nearly in line p. and f; about 14' apart; the p. one is of irregular outline; F; bM. The f. one is S; R; pB; bM.
4	1	Oct. 23, 1857. E. n. and s; B nucleus.
6	6	Oct. 12, 1855. E. n. and s; has a central nucleus and a * on edge, nf. the nucleus.
13	7	Resolvability not quite certain, irreg; R: nucleus.
15	6	Oct. 7, 1855. There are 6 or 8 *s of about 14–15 mag. and several smaller ones; I counted 7 knots, the 3 northern of which are the brightest; sketched. See fig. 1.
16	1	Sept. 19, 1857. S; R, or nearly so; and lbM. [Plate XXV.
17	2	Oct. 26, 1854. Several S; F. neb. visible at once in finder. 17 is R. and bM.
21	9	Spirality suspected, but more evidence wanting. Dec. 12, 1851. Nucleus; R. Oct. 22, 1857. I suspect outlying F. nebulosity, especially p. and f.
23 } 25 }	2	Nov. 4, 1850. 23 is R; and S. 25 has nucleus, and is E; 3 others near.
26	1	Dec. 6, 1850. R; F. nucleus; 40" br.
29	4	Aug. 21, 1852. Involves some *s, one of about 12th or 13th magnitude, E; vF.
30	1	Dec. 9, 1854. R; pL; bM.
32	2	Sept. 18, 1857. S; d. neb.; the n. one is E, sp. nf ; bM.
35	5	Sept. 1849. r; L; and rather F. Oct. 7, 1855. 3 *s in it.
37	2	Nov. 24, 1854. Not L; R; bM; a B. * close sp: r?
40	2	Dec. 6, 1850. sbM; another 18' distant.
42	2	Nov. 24, 1854. R; bM.
44	9	Oct. 16, 1855. vL; mE. np. by sf; sharp nucleus, for some distance round which, the neb. is B, and then *suddenly* decreases; there is a B. * np. the nucleus; and another involved in sf. end; another in p. border. Nov. 2, 1850. Spirality suspected.
45	1	Oct. 16, 1854. E. n. and s; many *s involved.
46	1	Dec. 7, 1857. s. of it is another neb.; E. nearly n. and s; 46 has B. centre; mE. n. and s, arms vF.
47	4	Nov. 3, 1855. Oval, and I think r; and has a * at np. edge.

Drawing not complete. The following are measures of some of the *s involved:—

		Pos.	Dist.
Dec. 18, 1851.	1.13	248°	6' 13"
	1.14	243	5 14
	1.15	260	5 51
	1.16	260	7 42
	1.17	241	7 30
	1.18	280	4 22
	1.19	301	7 35
Dec. 22, 1851.	6.20	219	3 46
	6.21	246	3 6
	6.22	234	5 1
	6.23	258	4 46
	6.24	271	4 20
	6.25	252	5 40
	6.26	295	4 34
Oct. 25, 1851.	1.N	83	2 2
	1.2	6	4 54
	1.3	309	5 1
Oct. 28, 1851.	1.4	1	7 25
	1.5	244	1 57
	1.6	254	6 13
Nov. 24, 1851.	1.7	8	7 32
	1.8	12	6 40
	1.9	11	5 46
	1.10	349	3 0
	1.11	345	2 10
	1.12	338	4 43

50 } 51 }

Number in Herschel's Catalogue.	Number of times observed.	Description.
53	1	Sept. 19, 1857. S; R; vF; bM.
54	2	Nov. 22, 1854. pB; vS; R.
59	3	Dec. 22, 1848. 3 neb. in line, 2 of them "novæ." Oct. 23, 1856. 1st is R; pB; bM; and has nucleus; 2nd bM; E, ✻ involved; 3rd F; 1E; bM.
60	1	Nov. 22, 1854. S; R; bM.
65	3	Sept. 18, 1857. S; pB. disc. in vF. haze of mottled neby.
69 } 70 }	7	Oct. 3, 1856. 69 is S; B; R; with B. nucleus; 70 is F; E. and patchy. I sometimes thought it was formed of two knots involved in F. neby; there appears to be a nebulous connexion between them all. Nov. 15, 1857. The silvered mirror shows the object brighter than before, but no new details: definition bad.
71	7	Suspect spirality; light unequal.
72	3	Oct. 26, 1854. a F. object with two nuclei.
78 } 79 }	4	Nov. 29, 1850. α is vlbM; β has stellar point or nucleus. I suspect δ to be a F. neb. Pos. Dist. αβ 219° 5′ 35″ αγ 315 1 8 αδ 81 0 44 Nov. 3, 1855. 3 neb. nearly in line, sp, nf; β is bM. and 1E. p. and f; α is R; bM; with a d. ✻ np, and is the largest of the 3; ε is S; F; R; δ is a ✻.
80	1	Oct. 3, 1856. pL; not vF. Its brightest part is a line running diagonally, and there is a knot at either end; believed to be a spiral.
84 } 85 } 86 }	4	Nov. 4, 1850. Pos. Dist. αβ 169° 2′ 19″ βγ 160 4 22 γγ′ 201 0 34 γδ 157 3 19 γε 176 5 32 εζ 199 1 41 θε* 79 4 55
87	3	Oct. 26, 1854. A d. neb., both S; R; bM.
89	8	A cl. with much unresolved neby.
90	1	lbM.
91 } 92 }	1	3 neb. in a triangle.
96	6	Oct. 26, 1854. Lenticular n. and s. Thought I saw a ✻ at times in centre (1¼-inch single lens); a lp. this is another vF. ray, np, sf, and which has no nucleus. Oct. 16, 1855. vF; E. n. and s; has nucleus; ✻ in n. end. Nov. 3, 1855. mE; pB. nucleus, and ✻ in n. end; np. this neb. is a ✻ of the 9th mag., and about the same distance p. this ✻ is another neb. vF; mE. Dec. 7, 1855. Seen as before; comp. neb. verified. Oct. 23, 1856. F. ray has nucleus and a ✻ in n. end. Sept. 18, 1857. E. n. and s; another vF. ray p, which is E. np. sf.
98	1	vF; R; S.
99	1	Oct. 3, 1856. S; F; R; bM; has nucleus.
103	3	Is n. of the 3rd of a group of 4 ✻s in line; 3 "novæ" near. Pos. Dist. Dec. 6, 1850. Aβ 28° 7′ 36″ Dec. 7, 1850. βδ 40 4 6 βε 81 9 19 Aβ 30 7 43
104	1	Oct. 23, 1856. 6 neb., all visible at once in finder eyepiece; 2 of them E., the others S; R; bM.
105	1	Dec. 11, 1854. vmE; bM (speculum dewed).
106 } 108 }	8	A variety of new nebulæ found, but observations too voluminous to transcribe.
112	6	Sketch made, but no interesting details. Nov. 30, 1850. vF, and p. a quadruple ✻. Oct. 23, 1851. 3 ✻s f. neb.; light unequal. Sept. 16, 1852. 2′ diameter; several ✻s in it; probably a F. cl.

* This should be, I think, θζ. A S. d. neb. suspected below at α′.

Number in Herschel's Catalogue.	Number of times observed.	Description.
113 114	2	Both have nuclei; "nova" near. Nov. 16, 1857. 113 is E. p. and f; * closely sp: 114 is R, with ragged edge and bM; "nova;" S; R; bM.
115 121	1	Oct. 3, 1826. The p. one is a pB. S. disc in F. outlying neby. The f. one is R; bM.
116	1	Dec. 18, 1851. s. end of neb. is like a brush or broom with a split.
118 120	2	4 neb. found, 2 have nuclei. 118 is S; R; 120 has 2 *s on np. edge; E. p. and f.
119	1	Dec. 9, 1854. pL; pB; bM to a nucleus.
123	2	Sept. 18, 1857. Rough sketch made; mE. np, sf; a F. triple * f.
128	3	Nov. 28, 1856. L; B; mE; B. nucleus. "Nova" f.

For item 131 (observed 27 times):

	Pos.	Dist.
Nov. 29, 1850. αβ	215°	0′ 51″
αγ	163	0 56
αδ	160	2 56
αε	178	3 07
αζ	192	3 44
αη	206	4 14
αθ	224	4 58
Dec. 27, 1850. αμ	147	5 34
αλ	179	5 56
αϰ	201	5 42
μν	143	6 28
α1	287	4 30
απ	341	6 45
Jan. 2, 1851. α2	5	5 18
αψ	357	4 42
α3	51	11 0
αρ	38	9 50
ατ	58	11 16
Dec. 23, 1851. αω	161	5 20
αα′	149	6 53
αβ′	172	6 32
αγ′	174	7 18
αφ	205	2 22

For previous observations see Transactions, Part II. 1850.
Sept. 13, 1850. Large spiral full of knots; to nf. is a S. neb. B, which on a very good night might appear attached to spiral, than which it is brighter. Oct. 11, 1850. Spiral arrangement not clearly seen. Nov. 27, 1850. Arms of spiral scarcely seen; fog. Nov. 30, 1850. Spiral form very indistinct; wind very high from s. Oct. 22, 1851. Viewed for drawing, I should not have seen the spiral arrangement had I not observed it before. Oct. 25, 1851. Neby. extends for several minutes all round, perhaps for half a degree in radius. Oct. 29, 1851. Observed for drawing. Dec. 14, 1851. Sketched. Dec. 26, 1851. Drawn. Dec. 7, 1855. This neb. reaches in length through at least a field and a half of finder eyepiece. Mr. Stoney's drawing leaves out a great deal of the neby. about the centre, and * suspected to left of centre of the trapezium of *s, perhaps others also. Nov. 15, 1857. There are 3 *s about the principal nucleus. Dec. 7, 1857. Carefully observed, with a view to a new sketch. Dec. 18, 1857. Carefully observed, and my sketch proceeded with. See fig. 10, Plate XXVI.

Number in Herschel's Catalogue.	Number of times observed.	Description.
132	1	Nov. 28, 1856. B; S; R. nucleus, a * p. and another n.
Nova.		Nov. 29, 1850. A S. neb. or cl. with 3 *s in it. Æ 1ʰ 26ᵐ. N.P.D. 60° 35′.
134 135	2	Oct. 26, 1854. Both S; R; B.
136		Sought for four times; not found.
142	8	Dec. 13, 1848. Rough sketch made. Spiral? Dec. 14, 1848. Confirmed last night's observation; feel confident it is a spiral. Oct. 24, 1851. Centre formed of *s: easily seen to be such; several *s through the neb.
143	1	Oct. 3, 1856. vS; F; R; bM; had a * close to n. edge.
147	2	Nov. 30, 1856. S; R; bM. to a nucleus.

Number in Herschel's Catalogue.	Number of times observed.	Description.
148	1	Dec. 11, 1854. S; R; bM. to a nucleus.
149	3	Nov. 30, 1856. Nucleus; E. np, sf.
150	6	A B. ray, with ✳ in s. edge, a little f. the nucleus.
151	2	Oct. 3, 1856. Long; vF; vlbM. A B. ✳ in p. edge.
156	3	Oct. 7, 1850. Light rather equable, a minute ✳ in the p. part. Nov. 24, 1851. E; a ✳ of 10th mag. nf. Sketched. Nov. 28, 1856. I see ✳s sparkling in it at times. See [fig. 2, Plate XXV.
157 159 }	4	A group of 5 neb.; others near.
161	2	Oval; ✳ in n. edge.
162		Looked for 8 times. Dec. 18, 1848. Found ✳ 7th or 8th mag. in place, but saw no nebulous atmosphere.
163	1	No nucleus. R; pF; bM.
164	1	Dec. 11, 1854. vB. nucleus.
165	1	Nov. 29, 1856. More E. than Herschel describes it; vbM.
169	1	A group of 5. Oct. 11, 1850.

	Pos.	Dist.
αβ	12°	0′ 30″
αγ	46	1 19
αδ	118	3 20
αε	296	1 59

| 172 173 } | 1 | { Nov. 24, 1854. d. neb.; the p. one is pB; R; bM. The f. one is smaller and fainter, and lbM. |
| 175 | 1 | Oct. 11, 1850. d. neb.; about 18′ nf. (169); nf. is a 3rd F. neb. |

	Pos.	Dist.
	171°	0′ 25″

181	7	Branches suspected several times, but not distinctly seen. Has a comp. neb. 5′ or 6′ s.
182	2	Oct. 23, 1857. S; pB; R; bM.
183	2	Dec. 7, 1850. Nucleus E. np, sf.
188	1	Nov. 30, 1856. S; F; R; lbM.
190	1	Nov. 22, 1854. eeF; E; no nucleus. A ✳ 10th mag. p: several S. ✳s near.
193	2	Dec. 18, 1856. bM. to a nucleus. E. sp, nf; S. ✳ in s. end.
194	1	Some ✳s in it.
195	2	Nov. 30, 1856. E. sp, nf; a F. ✳ follows closely; there is another F. ✳ in n. edge.
197	4	Dec. 23, 1851. I suspect a F. appendage f; a d. ✳ f.
198	1	Nov. 24, 1854. vF; mE; almost lenticular.
199 200 201 202 }	2	Nov. 29, 1856. All are S; R; bM.
205	1	Nov. 28, 1856. S; R; lbM; a ✳ in centre.
207 212 }		{ Sept. 13, 1850. Between the 2 cls. there is a red ✳ nearer the 2nd, and 2 more red ✳s f. 2nd cl. of 8th or 9th mag.
208 210 }	14	{ Nov. 3, 1855. A dark space running along s. side of nucleus of 210, and (Nov. 5, 1850) ✳ in sf. extremity; r. Both have S. comp. nebs. s; 208 is E; ✳ close f. centre.
212	1	No description.
213	2	Nov. 30, 1856. vS; eeF; R; vlbM.
215	2	Oct. 29, 1851. Nucleus; 5′ n. of d✳.
216	1	Sept. 17, 1852. ✳ in edge; perhaps like a snowdrop.
217	4	Sept. 14, 1850. Oval; mbM; pB; 50″ by 70″.
218	7	Dec. 27, 1850. Pos. of chink 19°. Dec. 26, 1851.

	Pos.	Dist.
αβ	44°	1′ 57″
αγ	198	4 38
αδ	199	5 48
αε	211	2 38
ray	23	10 29

| 219 | 1 | Nov. 28, 1856. d. neb.; components unite at p. end. The s. one is L, E, and gbM. The n. one is more E. and fainter, and also bM. |
| 221 | 2 | Oct. 23, 1857. L. but eF; mottled; ✳s in it, especially one closely n. of centre. |

Number in Herschel's Catalogue.	Number of times observed.	Description.
222	1	Sept. 14, 1850. 3' by 50''; rather F. dash of light; a conspicuous * nf. the M. outside edge.
223 } 224 }	2	Nov. 29, 1856. 223 is pL; B; vbM; R? It seems to have some F. *mottled* neby. about it. 224 is vF; pL; vlbM.
226	3	Oct. 16, 1855. Oval; no nucleus; light pretty equable; major axis np, sf; clearly r. I can at moments see some of its *s. A B. * at s. edge.
229	1	Nothing particular.
230	2	Nov. 24, 1851. Brightest part near p. edge; E. nnf. ssp; d. * n, to which neb. does not reach.
231 } 233 } 234 } 237 }	2	Oct. 11, 1850. αβ Pos. 83° Dist. 3' 43'' αγ 22 1 59 αδ 40 2 31 Another about 12' sf.
232	9	Oct. 12, 1855. Sketched; r. See fig. 3, Plate XXV.
238	3	Dec. 12, 1848. bM. nearly to nucleus.
241	8	Sketched twice, Dec. 11, 1854, and Nov. 23, 1857. See fig. 4, Plate XXV.
242	8	Dec. 27, 1850. αN Pos. 73° Dist. 0' 54'' αγ 81 3 19 αβ 331 0 29 αδ 282 1 39 Sketched four times. See fig. 5, Plate XXV.
244	1	Nov. 28, 1856. Patchy; pL; mbM.
246	2	Spirality suspected; E; gbM.
247	1	vS; R; F; bM.
254	1	gbM.
255	4	Dec. 27, 1856. r; has 3 *s in edge, and I think I see one just p. the nucleus.
256	1	E.
257	6	Jan. 2, 1851. 5 knots; the p. one is d. neb. Dec. 26, 1851. A ruddy * of 10th mag. p. 16'.
258	4	Nov. 30, 1850. A F. dash of light nearly p. and f; the n. edge is the best defined.
262	12	Sketched 4. Dec. 22, 1848. A blue spiral. Jan. 14, 1849. Spiral. Oct. 29, 1851. The central part is flatter on the f. side. Nov. 24, 1851. The central part is, I am nearly quite sure, spiral, sketched. Jan. 13, 1852. Spiral form of centre seen. Nov. 29, 1856. Details of drawing seen very well. Jan. 10, 1858. I can see nothing more than is given in the sketch, which appears to me correct, though perhaps it defines too well the edges of the B. central disc. See fig. 6, Plate XXV.
263	1	Dec. 7, 1850. R. nucleus.
264	6	Nov. 23, 1848. A curious object with dark spaces. Oct. 10, 1850. r. Oct. 16, 1855. Fine oval neb; has nucleus; light mottled; sometimes I thought I saw a dark bay n. of nucleus; certainly the neb. is brighter along the n. and nf. side than in the part intervening between that and the nucleus. Dec. 6, 1855. Previous suspicion as to direction and existence of dark streak confirmed; the nucleus and n. edge of neb. both seem r.
265	2	Jan. 7, 1849. F. patch, 2 *s perhaps, p. middle.
266	3	Jan. 14, 1849. vF; lE.
269	1	Very badly seen.
271	3	d., with another knot near.
273	2	Dec. 7, 1857. F; S; R; lbM.
275	6	Appendage suspected; E. n. and s; bM.
276	2	Dec. 7, 1857. vS; R; F; a S.* close sp.
277	3	Dec. 9, 1857. pB; oval; has a B. central nucleus; about 4' n. is a F. E. neb. containing *s.
279	1	Dec. 11, 1854. Has a B. * sp. the nucleus.
280	3	Badly seen.
282	5	Oct. 10, 1850. BM; r.
285	6	Nov. 29, 1851. E. p. and f; bM.
286	8	Oct. 30, 1851. E. p. and f; bM; between this neb. and 282 there are very few *s.
287		cl.

Number in Herschel's Catalogue.	Number of times observed.	Description.
288	1	S; R; vF; bM.
289	3	Dec. 8, 1850. Double; γ ✳ of 9th mag; α is 289, and has a F. nucleus; β "nova." Pos. Dist. αγ 2° 2′ 53″ γβ 152 2 08 γδ 103 1 54
290	5	Nov. 23, 1848. Coarse, cl. strongly honey-combed. Would probably look annular with eccentric eyehole if it were far enough to be a neb. Nov. 21, 1851. The honey-combed appearance is caused by the disposition of the brighter ✳s; no spiral arrangement.
292	3	Jan. 17, 1855. r. Nov. 28, 1856. Edge ragged.
293	4	Dec. 16, 1848. A multitude of nebs. knots in the neighbourhood, principally p; counted 15; many more. Dec. 8, 1855. One of them F; has a ✳ close sf. and looks like a snowdrop.
294 } 295	1	Nov. 24, 1854. Two S. R. neb.; both bM.
296	5	Dec. 19, 1848. gbM; E. sp. nf.
297 } 298	10	2 "novæ." Oct. 10, 1850. Pos. Dist. βα 143° 1′ 35″ βγ 11 6 1 Nov. 2, 1850. Another 8′ n. of γ.
299	2	Dec. 16, 1854. vF; lbM. to a nuc.; mE. np, sf.
301		Scattered cl.
302 } 303	4	302, r. 303, mottled.
304	1	Jan. 17, 1855. d. neb.; both vS. and bM.
305 } 306 } 307	8	Dec. 26, 1856. The p. one is vF. and light mottled. Oct. 7, 1850. 1st appears divided, and preceding part has a minute ✳. Jan. 22, 1851. f. the 3rd; 14′ is "nova."
308	1	Oct. 31, 1856. A fine d. ✳ in a loose cl.
309	2	Oct. 26, 1854. S; R; bM.
310		Cluster.
311	9	Sketched five times. Jan. 13, 1852. New spiral of an annular form round the ✳, which is central. Brightest part is sf. the ✳; spirality is vF; but I have no doubt of its existence. Oct. 7, 1855. Annular, but with a break in s. side of annulus, or perhaps spiral. Oct. 31, 1856. I feel certain of a dark space nearly p. the central ✳, but the shape of the whole is only conjectural; there is a ✳ plain np. the neb. Dec. 7, 1857. Not vF, and the break in the s. side of the ring of neb. quite easily seen; *between* this ring and the central ✳ is not black, but filled with more F. neby. Jan. 9, 1858. Observed for a sketch; last observation correct as to shape; the brightest part is sf, and the next brightest is on the opposite side, and with ½-inch single lens the whole annulus has a mottled look. Jan. 13, 1858. The whole edge was ragged and irregular, and the whole neb. much mottled. See fig. 7, Plate XXV.
313	2	Dec. 16, 1854. vF; S; R; lbM.
315	16	Sketched five times. Jan. 2, 1851. Dark space sf. neb.; though I did not see the F. neby. beyond the channel, I conjecture that it exists and fades off imperceptibly, somewhat like the drawing. Jan. 22, 1851. Observations of Jan. 2nd confirmed; F. neby. seen; B. part r. Nov. 29, 1851. Last season's observations confirmed as to shape. Dec. 22, 1851. Previous observations confirmed. Jan. 13, 1852. ✳ is d; ξ is the angle of the brightest part. See fig. 8, Plate XXV. Pos. Dist. αβ 57° 0′ 55″ αγ 204 3 50 αδ 199 4 1 α✳ 315 2 23 αξ 223 1 42

Number in Herschel's Catalogue.	Number of times observed.	Description.
316 317 318	7	Dec. 5, 1850. α and γ are bM; γ is about 10' nf. α, and has a brush-like elongation (see 242) at each end. Oct. 7, 1850. αβ Pos. 77° Dist. 0' 56" Dec. 5, 1850. αβ 75 0 58
319	2	3 "novæ" near.
321	3	Oct. 26, 1854. Has a ✳ at n. extremity; E. np. by sf; Herschel's d. ✳ nf. is triple. Jan. 15, 1855. The conspicuous ✳ involved in n. end of neb. has a F. comp; nf. itself very distinct with 1½-inch single lens.
320 322	1	3 neb. nearly in a line; one "nova"; 1st pL; F; R; 2nd pF; R; 3rd dull nucleus.
327	10	Sketched twice. Appears to be a spiral, but evidence not quite satisfactory. See fig. 9, [Plate XXV.
331	1	Nov. 29, 1856. 2 ✳s near edge; vS; irreg. R.
334	6	4 neb. (3 "novæ"); one of them is E.
336	6	Jan. 13, 1858. B. centre; F. neby. stretches out a long way, involving a minute ✳ p.
338		Oct. 26, 1854. A group of a few ✳s.
339	1	2 nebs. knots.
340	2	"nova" near.
343		Looked for seven times. Not found.
347	15	Dec. 11, 1850. A S. comp. p. and a d. ✳ n. Jan. 10, 1858. Looks like a F. haze enveloping 3 ✳s.
349		Large loose cl.
352	2	Close d. neb.
354		Dec. 28, 1856. Neat little cl; its centre consists of about 40 or 50 ✳s; the outlying ✳s are arranged in curved branches.
355	5	Nov. 29, 1848. Saw a multitude of ✳s and some unresolved neb.
356		Looked for four times; not found, but nights bad.

357	19	Sketch not quite satisfactory.
		
		Nov. 29, 1851. γδ Pos. 351° Dist. 0' 47"ᵃ
		    γζ 52 2 10
		    γη 40 2 16
		    γθ 70 2 53
		    γε 110 3 19
		    γι 104 3 37
		Jan. 12, 1852. δγ 348° 1' 36".

358		Coarse cl.
359	1	Dec. 28, 1856. Looks like a ✳ in vF. neb. atmosphere. 1E. p. and f.
360	43	(Neb. of Orion.) Account of detailed observations postponed, as in 50 and 357.
361	11	Observations recorded in the 'Transactions' for 1850 fully confirmed.
363	6	Nov. 30, 1850. The luminous appearance extends about 15' all round the ✳.
365	3	Oct. 23, 1851. r: I strongly suspect it is annular.
368	8	Feb. 9, 1852. Spiral arrangement sufficiently seen to confirm former observations. Jan. 9, 1856. Appears in finder a B. oval neb., with n. and nf. edges brightest and best defined, and sp. edge fading away gradually; with higher power there is seen a decided darkness at and between the ✳s, and I can confirm previous observations as to the curve formed by the brightest part of neb. Dec. 26, 1856. Nebulosity easily traced as in preceding sketch.
370	1	Jan. 21, 1857. r? suspect ✳ in centre.
373	3	Nov. 30, 1850. Same appearance as ε Orionis, but very much fainter.
375		Jan. 17, 1855. A pretty close cl. of S. ✳s, followed by four or five B. ✳s.
378 381 383	7	Dec. 11, 1850. I saw no nebs. round 378; sf. about 20' is a triple ✳, the middle one of which is pretty strongly nebs.; about 36' f. (a little n.) is a d. ✳, whose brighter component is nebs.; 65' f. 378 is a S. neb. with nucleus or stellar point.
384	1	No description.
385	2	A few B. ✳s; scattered.
389		Dec. 28, 1856. Very loose cl.

ᵃ Note by observer: "I have reason to believe that the distance of γδ is incorrect."

Number in Herschel's Catalogue.	Number of times observed.	Description.
390		Jan. 20, 1857. A close irregular cl. of vS. ✳s; figure as sketched, one ✳ rather brighter than the rest; forms as it were a nucleus, round which the others are grouped, but principally np. side of it.
393	15	Sketched three times. Feb. 28, 1850. S. ✳ near a L. one; the L. ✳ f. the neb. has a comp; this, No. 393, is an enormous neby, which I traced f. and n. of it to a great distance, some degrees. It narrows at times to a band across the finding-eyepiece of about 6' or 8'. I fancied the number of L. ✳s was greater in it than in the neighbourhood; I am certain the number of S. ✳s is much less. In a small space, taken at random in its neighbourhood, I reckoned upwards of 20 S. ✳s. In a similar space in it, taken at random, but 3. See fig. 11, Plate XXVII.
399	11	Feb. 22, 1851. 2 ✳s in p. part of the neb. Nothing additional to what is in the 'Transactions' for 1850.
401	9	No neby. found, and only a few ✳s arranged in pairs; no cl. Has there been a change here?
403		R; with rays.
404		Jan. 10, 1856. S. cl. of S. ✳s; oblong nearly p. and f.
406 407	12	The southern one has nucleus.
408	1	
409 410	7	Dec. 8, 1850. 5 nebulous knots. Pos. Dist. αγ 344° 2' 32" αβ 323 1 46 αδ 3 5 08 αε 30 6 11
411		Nov. 25, 1851. A coarse B. cl.
413		Jan. 20, 1857. Pretty cl. of pB. ✳s; centre nearly R.
415		Jan. 8, 1851. A poor cl.
421	6	Nov. 23, 1851. v. close d. neb. below 4 ✳s. See fig. 12, Plate XXVII.
425		Feb. 13, 1852. Coarse cl.
424 426	3	Both bM.
427		v. loose cl.
428	2	Feb. 1, 1856. vF. fan-shaped neb. involving 3 ✳s.
430	2	Jan. 31, 1851. Several knots around; 430 is E. np, sf.
431	2	Jan. 18, 1855. S. ✳ in s. edge. Jan. 25, 1857. r.?
434	10	✳ in f. edge; r.
439		Jan. 9, 1856. Loose cl; irreg; R.
443		Jan. 30, 1856. About 25 or 30 ✳s of a curious shape.
444	15	Nov. 23, 1851. S. ✳ in f. end of np. appendage, also one nf. the neb. about 40"; nothing additional to description in 'Transactions' for 1850.
446 447 448 449	5	Feb. 26, 1851. Pos. Dist. αβ[a] 222° 3' 41" βγ 282 5 37 δε 291 3 49 εζ 267 6 47
450	25	Although 21 observations have been made since the sketch appeared in the 'Transactions' of 1850, nothing additional has been discovered, except that the outer luminous ring is of unequal brightness.
453	1	Jan. 21, 1857. S. ✳ close to n. edge.
454		Neat little cl. of vS. ✳s. It looks in finder like a r. neb.
456	11	Edge filamentous; r; vlbM.
457	5	Edge filamentous; r; looks like a globular cl.
458	1	No description.
464	16	Dec. 8, 1850. Dark space more eccentric than in the drawing in the 'Transactions' for 1850. The larger of the 2 ✳s is in the dark space, the other s. of it in the neb; a 3rd ✳ close nf.
465	1	Jan. 11, 1856. pF; bM; 1E?

[a] α, β, γ and δ, ε, ζ are distinct figures.

Number in Herschel's Catalogue.	Number of times observed.	Description.
468	5	A group of 6 *s ; no neby.
469 470	1	Feb. 20, 1851. A great many knots ; reckoned 10 in a line nearly p. and f.
471	1	Jan. 9, 1856. d. neb.
473	1	Feb. 1, 1856. F. ray, with pB. nucleus; np. this is another neb. vF; E; with * near nucleus.
476	2	Jan. 12, 1855. A F. * p. and a nebulous knot f.
477	1	Another near, both F. S.
478	16	Jan. 20, 1855. I see 2 *s in p. edge with ½-inch single lens. The smaller component of a double * touches f. edge; light mottled. On several occasions spirality suspected, and rough sketch made.
480	Frequently.	Several observers have fancied that the *s exhibit some approach to a spiral arrangement, with cellular centre. No unresolved neby.
481	1	Jan. 20, 1857. * in n. edge; centre r?
482	1	Nucleus; vF; R.
483 484	3	Jan. 31, 1851. 2 others near 483. Feb. 26, 1851. αβ 242° 2′ 48″ / αγ 319 5 36
486	3	Feb. 16, 1855. L; vF; lE; light mottled; suspect dark spaces round the centre; F. stellar point or nucleus; several *s in edge and in it. Rough sketch made.
487	1	Jan. 25, 1857. Light mottled; B. * n; a F. * close nf. edge.
489	10	Feb. 14, 1857. Certainly a * in centre or nucleus, and neby. projecting to sp. side, but eF. Mar. 10, 1858. (Definition very good.) Nucleus stellar; the brightest part of the neb. looks r. It is pL. and mottled; suspected spiral.
491	15	Sketched 6 times. Spiral. Jan. 29, 1856. Very well seen; previous observations confirmed. I have no doubt the neb. is a spiral. The f. half of neb. is the more difficult to see well. Mar. 10, 1858. Well seen; the whole neb. looks vB. and sparkling; part is clearly r; my former conjectures as to its shape confirmed. I used the highest single lenses. Mar. 11. 1858. (Definition very good.) Observed with same results as on last night.
492	6	But never well seen.
494	1	Jan. 20, 1857. vF; E; nearly n. and s; has a sharp pB. nucleus.
495	2	Feb. 28, 1851. Centre r; E. n. and s.
496		Jan. 31, 1851. Coarse cl; lanes and openings without any *s whatever.
497	4	In the centre of a triangle formed by 3 minute *s; nucleus.
498	2	Feb. 22, 1857. 2 *s on np. edge; S; F; R; bM.
499	3	Nucleus; F. * in p. edge; S; vF; R.
504	2	r; * in f. edge.
505	10	Jan. 27, 1852. r. Feb. 9, 1855. Centre suddenly B; irreg. R.
506	4	Jan. 17, 1855. Centre suddenly B. with outlying F. neby, which involves a * nnf.
507 508	3	Feb. 9, 1850. A fine object; 3 neb. forming a triangle; one B, another pB; the third the last degree of faintness.
510	1	4 neb. here. The f. one is E. and has nuc; the others are S. and F.
512	5	Jan. 10, 1856. Not vS. but vvF. and flickering. Feb. 1, 1856. Nucleus and * close to s. side of it, and two very indistinct branches of neby. From the tenour of the observations no doubt it is a spiral; the twist of the branches fully confirmed.
513	9	Nov. 30, 1850. S. * in its nf. edge, perhaps not connected with the neb. The neb. had a brush-like appearance. Feb. 1, 1851. Dark space f. the * between neb. and *, like the " snowdrop " neb. (see fig. 10, 'Transactions' for 1850).
514	20	Dark space suspected in centre, but never fully confirmed. Remarkable for extreme paucity of *s in neighbourhood.
518	11	Nucleus surrounded with L. F. neby.
519	6	Jan. 30, 1856. 4 *s in vF. neby.
521	1	Feb. 23, 1857. E; np; sf; mbM.
522	4	Light *not* equable; stellar nucleus?; * in n. edge.
526 527	3	Feb. 9, 1855. Very close, almost touching. 526 is mbM; 527 is smaller, and lbM. Sketched.
529	7	March 9, 1852. e. close; d. neb. Jan. 20, 1857. These two are equal in size, and enveloped in F. haze of neby.

Number in Herschel's Catalogue.	Number of times observed.	Description.
530	7	B. nucleus, surrounded by very extensive neby.
531		Coarse cl.
532	7	Dec. 29, 1851. vL. lenticular ray, slightly concave towards np. direction; gvmbM; perhaps 10' long. March 1, 1854. Uncertain whether nucleus is stellar. Query, parallel dark lines exterior to nucleus as in Andromeda. March 8, 1858. * on np. edge is d.
533	1	March 11, 1858. 4 neb. here, nearly in line p. and f.
535	9	d. neb. surrounded with F. neby.
536	6	6 knots in the immediate neighbourhood, two of which have * in their edges.
537	2	* with fan-shaped neb, very like Herschel's fig. A 2nd F. star involved in the neby.
538	1	March 1, 1856. F. bM.
540	8	r; d. * in s. extremity; nucleus.
542	1	Feb. 19, 1855. vvF. nucleus; r.
549	1	Feb. 18, 1855. d. neb.
550	1	March 12, 1852. An amorphous mass of neby. of uneven character; E. p. and f.
551	3	Jan. 20, 1857. lE. p. and f; and vlbM.
553	4	Feb. 16, 1858. Nucleus; pB; E. nearly n. and s.
555	3	Dec. 29, 1851. A B. ray like 242.
556 559 561	2	4 neb, one of them vvF. and one E.
557	1	Feb. 23, 1857. vF; lE; lbM.
562	4	vF. ray; np, sf.
563	7	Feb. 16, 1858. Mottled, and suspect spiral; r. March 11, 1858. B. * close f.
564	2	March 26, 1851. vbM.
566 567	1	March 13, 1850. A third, and eF. neb. found.
569	2	lE.
574	1	Jan. 8, 1851. * in edge; R; S.
575	2	Feb. 9, 1855. 2 neb. found; both F. and lbM.
580	2	March 15, 1855. E. nearly n. and s.; has a * touching its nf. edge, and is mbM.

581 582	5	There are here 15 knots. The positions of six of them were taken.
		
		                   Pos. Dist.
		March 26, 1851. αβ 226° 0' 25"
		                   αγ 237 1 12
		                   αδ 263 5 9
		                   αε 125 4 13
		                   εζ 120 8 8
		March 14, 1850. αβ 235 about 0 30
		                   βγ 245 about 0 45

584	5	The neb. involves one of Herschel's *s.
587	1	Mar. 15, 1855. Has a F. knot close np.
588	1	Nucleus.
589 591	4	3 found; 2 of them E. and lbM.
592	2	*s in its edges, and suddenly condensed in the centre.
593	5	Feb. 14, 1855. Stellar points in outlying F. neby, especially two, which I can plainly see with the ½-inch single lens; sbM.
594	1	Mar. 9, 1858. E. neb. between 2 *s.
597 598	3	Feb. 22, 1857. Fine d. neb, both mbM. and both E, especially the f. one, which seems to have a bend at α. Query, a vF. neb. at β? Mar. 18, 1857. All the particulars of my last observation fully confirmed. "Nova" at β seen.
600	3	Feb. 19, 1855. pF; R; bM. to nucleus.
604	24	(18 times since 1850.) Nothing additional, except 3 *s as in diagram. Mar. 9, 1858. Very well seen; central nucleus looks r. A * suspected at α, and one or more in the F. neby. at β; and a * seen at times quite steadily at γ. I employed the inch and ½-inch single lenses. March 11, 1858. Seen as well as on last observation. I have now verified the 3 *s which I then noticed.
610	4	Mar. 24, 1857. Much mottled. Mar. 11, 1858. Has a d. * in it.

Number in Herschel's Catalogue.	Number of times observed.	Description.
613	1	Feb. 18, 1855. vF; R; mottled?; ∗ in n. edge.
622 624 627	8	Feb. 1, 1856. 622 has nucleus, and is mE; its light is very unequal, and I suspect one dark lane running throughout its length; s. of nucleus.
626	2	Jan. 30, 1856. pL; vF; R; vgbM.
630	1	Mar. 5, 1851. S; lE; vgbM.
634 636	4	Feb. 19, 1855. The p. one is d; its comp. being immediately p. it, and lE. sp. by nf.
635 637	2	Mar. 9, 1852. n. one has a mottled appearance.
638	1	Jan. 25, 1851. r; 5 "novæ" near the most distant 11'.
639	10	Jan. 24, 1851. r; a S. ∗ near the middle, and another f. lenticular. Mar. 20, 1851. Patch and ∗ in p. end. At 54° 46' N.P.D. 9ʰ 36' Æ } ± A scarlet ∗ of 18th mag.
640 641	1	Several knots near.
642	1	3 "novæ" near.
644 647	2	Feb. 26, 1851. p. one eF.
650	2	psbM.
652	2	Mar. 10, 1852. L. thin F. ray.
656	4	Mar. 15, 1855. Appendage to sp. edge, or rather a twist in that end towards the north.
657	12	vF; seems to have a split in f. end.
659	2	Nucleus or ∗ in M.
660	1	Nucleus or ∗ in M; light mottled; a S. ∗ nf.
661	1	Mar. 21, 1854. eeF. with B. centre; E, principally on f. side.
663	3	Jan. 10, 1856. Lent.; vbM; has a ∗ np. Query, a break in the neb. just p. the nucleus?
665	1	Mar. 18, 1857. Found here a ∗ with vF. neby. nf. it.
667	3	Mar. 24, 1857. pF; S; R; bM.
668	6	Mar. 11, 1848. Fine ray, with vB. nucleus.
671	6	Mar. 12, 1852. E. p. and f.
675	4	S. ∗ sp. edge.
677	1	Mar. 30, 1854. vF.
678	1	Mar. 27, 1854. 3 neb; the p. one vS. About 4' f. is a S. lent. ray running nf, sp, and s. of this latter is another neb. about 5' distant; R; both r.
682	8	Jan. 16, 1850. A F. spiral. Mar. 20, 1854. A F. ∗ immediately f.; spiral left-handed; very faintly seen; night bad.
684 685	5	685 seems like 393, but instead of the ∗ having an approach to a nucleus. Jan. 30, 1856. About 5' sp. 684 is a vvF. ray, extending n. and s. 684 has a B. central nucleus, with a sensible disk.
688	4	Suspected spiral.
689	7	Spiral. Feb. 1, 1856. The neby. connecting the three principal knots is vvF, but no doubt of its existence. Sketch made. See fig. 13, Plate XXVII.
692 693	10	March 15, 1850. 4 neb. here. α is 692, and β is 693. Pos. Dist. αβ 51° 5' 50" αγ 302 5 10 αδ 217 10 47 αβ 53 5 52 According to Herschel, the distance from 692 to 693 is 4'; this should be carefully looked after. Mar. 22, 1857. Sketched; α and γ r; nucleus of δ appears eccentric. See fig. 14, Plate XXVII.
695	5	March 3, 1850. Probably a F. spiral. March 24, 1857. ∗ in f. end; dark spaces throughout its length.
696 699 700	1	Feb. 16, 1855. They form an obtuse-angled triangle; the p. one is accompanied by 2 minute ∗s, one n. and the other nf; the next has also a minute ∗ as a comp. nf.
698	1	vF.

Number in Herschel's Catalogue.	Number of times observed.	Description.
705	1	March 3, 1851. d. *, with neb. to n.
706		Looked for 5 times; not found.
710	5	March 11, 1858. 2 F. patches of neby. (of which one has nucleus); they form with a * an obtuse-angled triangle, the intervening space being filled with F. neby. of a mottled character.
711	4	Feb. 18, 1852. pB; bM; E. sp, nf.
713	1	bM.
714	2	March 20, 1854. Dark spaces suspected. Feb. 9, 1855. Has a suddenly B. centre; vmE.
718	1	Has a * closely sff.
719	2	Rather lenticular.
720	2	March 18, 1857. sp. edge is F, and not so sharp as the rest.
721	2	March 18, 1852. R; nucleus.
724	6	March 8, 1858. There is a B. streak, in which I certainly see *s sparkling, projecting a little from the edge of neb; the neb. is much mottled, and has a stellar nucl.
727	1	March 11, 1858. F; R; bM.
728	5	Jan. 10, 1856. I think the nucleus is not quite central.
731	12	March 5, 1848. Spiral arrangement well seen. March 11, 1848. Very cold; very windy; air steady; definition excellent; mirror bore a power of 700 with great precision; telescope as steady as a rock, although wind so high. Nebula well resolved into * points. Saw a broad band at the bottom distinctly, and 2 at the top. March 28, 1848. Resolved by a power of 800, although night hazy. March 17, 1849. Like cl. in Hercules; dark spaces in B. part.
732	1	Between 2 *s, one of which seemed connected with the nebulæ.
735	1	Nucleus or * in centre; S; R.
737	4	March 18, 1857. Mottled; suspect 2nd nucleus.
739	8	Jan. 27, 1852. vF. spiral with B. centre; S. * sf. centre involved; two others f.
743 } 749 }	14	Both have B. L. centres enveloped in F. neby; much mottled. 743 sketched roughly twice. 749 sketched roughly four times. They are both represented as spirals, though the details are vF.
748 } 751 } 753 } 754 }	3	Mar. 23, 1851. The triple neb. is probably a spiral; dark spaces in it.
750	4	Feb. 1, 1856. lE; pB; mbM.
755	3	Feb. 23, 1857. mE. n. and s; bM.
756	3	B. streak through it suspected.
757 } 758 } 761 }	3	March 17, 1849. { 757 vB; L; R. / 758 vB; R. / 761 E; pB.
765 } 766 }	9	Feb. 9, 1855. 765 is, I think, a spiral, with *left-handed* twist; immediately f. is 766, which is B. and well-defined. I suspect F. neby, extending from 765 and running up through the other nebulæ. Feb. 14, 1855. Seen as before. In 765 the curve to the left is brightest near its extremity. Feb. 16, 1855. Certainly F. neby. extends between the two, as before suspected. Jan. 10, 1856. Nothing to add to former observations. Mar. 19, 1857. Observed to compare sketch. See fig. 15, Plate XXVII.
768	1	Nucleus.
772	1	Another neb. n. 3' dist.
773 } 775 }	7	Both mottled.
774	1	Sharp nucleus; * in nf. edge.
777	3	Mar. 29, 1856. A * in s. edge, and a F. one in f. edge; 2 knots in n. edge. I think it r.
778 } 779 } 782 }	3	Feb. 9, 1855. Three in a line; the middle one is vB. and lenticular, and has the larger * of a d. * involved in f. end.
783	1	vF; lbM.
785 } 787 }	3	Mar. 5, 1851. At sp. edge of 787 a ring suspected, within which a dark band, α then B. part. Mar. 30, 1856. 785 is E. sp. by nf, and its brightest part is nearest the p. end; also a * in nf. edge. 787 is very curious. A R. bright nucleus, which is eccentric, and a dark curved passage sp. the nucleus, as in sketch. The neby. outside this dark place runs up perhaps to the streak marked α, which is vF, but of its existence I have no doubt.

Number in Herschel's Catalogue.	Number of times observed.	Description.
786	1	Stellar pt. or nucleus E.
788	6	Jan. 1850. Probably very remarkable; bad night. Feb. 1, 1851. f. division the brighter. Mar. 3, 1851. p. division pretty well seen. Mar. 8, 1856. mE; certainly dark spaces on each side of nucleus, but not well seen; that on f. side is the more distinct. Sketched roughly 3 times.
789	1	Jan. 21, 1855. pL; considerably E; BM, but no nucleus.
790 } 791 }	2	{ Mar. 28, 1856. About 3′ apart; both F. and of nearly equable light. The n. one is a long narrow ray np, sf; the other is oval sp, nf.
793	1	A S. comp. dist. about 5′ or 6′.
804	3	April 9, 1852. I suspect a dark curved passage sp. centre. Mar. 15, 1855. Light mottled; I suspect a knot in p. and one in f. edge. Has a spiral appearance.
805	6	Mar. 12, 1855. Has a sharp B. R. nucleus in a disc of F. mottled neby.
806	2	Mar. 17, 1855. Oval; major axis nearly p. and f; nucleus vB.
810 } 815 }	4	{ Feb. 22, 1857. mE; B. nucleus; arms F; patchy. Mar. 23, 1857. pL; nucleus vB, and has a sensible disc; arms vF. and patchy. 815 is F, nearly R, lbM.
811	1	R; gbM.
812	1	Query, is there a F. ring round it?
813	5	Mar. 1, 1854. Query, an oval spiral?
814	2	April 13, 1852. Neb. does not appear to reach the *.
818	4	Very like H. 2172. See figure. Mar. 29, 1856. The nucleus projects into the space along sp. edge; outside this dark space there is F. neby, which I see joining the neb. at n. A·F. * at the opposite extremity.
831	3	April 3, 1851. Light mottled; vBM; knot in p. branch.
838	42	April 13, 1850. But one * seen. Feb. 1, 1851. 2nd * not seen; sky milky. March 3, 1851. 2nd * not seen; sky milky. March 5, 1851. 2nd * not seen. March 7, 1851. 2nd * not seen. April 3, 1851. 1st * only seen. Jan. 27, 1852. Only one * seen. March 12, 1852. Only one * seen. March 13, 1852. Only one * seen. March 20, 1854. 2nd * not seen, nor any of the F. details. March 30, 1856. 2nd * not seen, nor minute details. March 24, 1857. 2nd * not visible. March 8, 1858. 2nd * not visible, nor minute details. N.B. The 2nd * has not been seen since March 9, 1850.
840	5	March 17, 1855. mE. p. and f; vB. centre; the n. edge of central part seems sharpest, and outside it again I think there is F. neby; * in f. edge. A rough sketch represents it like 2172.
841	3	Suspected spiral, but a vF. object.
843 } 844 845 846 }	4	844 is a B. nebulous disc in a F. oval neby.
847	1	Mar. 19, 1852. E. np, sf; vB. centre.
848	3	Feb. 18, 1852. vbM; lE.
849	1	Mar. 29, 1856. S; pB; R; mbM.
851	1	" Nova " near; both are S; F; lbM; and 851 has nucleus.
854	10	Mar. 31, 1848. Curious neb. with B. nucleus at left; a little above and towards the right is a streak; spiral; resolved very well about the nucleus, but no other part. From the right, and apparently springing from the nucleus, a vF. portion of neb. extends for nearly 15′, gradually melting away. Apr. 3, 1848. Observed with the same results as on March 31st. April 17, 1849. 2 *s near nucleus, one sp, the other sf. it. Feb. 25, 1854. Suspect dark spaces on either side of nucleus. Mar. 1, 1854. Neb. mottled; p. observation confirmed.
856	2	2 neb. found; the p. one has a sudden vB. nucleus, and is lE. np, sf; the other is about 15′ f; S; R; pF; vlbM.
857	4	Suspected darkness on either side of nucleus; E. See fig. 16, Plate XXVII.
858	4	Apr. 15, 1852. R. disc, BM, with vF. neby. round it of mottled character; probably it will be seen as spiral on a fine night. Mar. 30, 1856. Spiral with, I think, two arms, thus: these arms are broken and of unequal light; there are B. patches at α, β, and γ respectively; a F. * p. at δ. Apr. 6, 1856. Seen as spiral. The f. branch comes down past the other, doubtless over it as at α, and seems to originate from the p. side of nucleus. Mar. 24, 1857. The spiral arms are eeF; but there is no doubt of their existence as described in previous observations.

Number in Herschel's Catalogue.	Number of times observed.	Description.
859	2	Apr. 1, 1848. pB; very long.
860	1	Apr. 9, 1852. I see nothing but a F. neb. 60″ near some *s of 8th and 9th mag.
865	1	bM.
866 } 869 }	1 {	Apr. 13, 1852. Large neb. is BM. It has a knot in sp. end, and a dark curved passage on p. and n. sides of centre; spiral. Small neb. f. has a S. * immediately s. of it.
875	6	Sketched 4 times. Feb. 19, 1855. 3 *s in it; there is a mass of neby. f. the brightest part, with condensed portions through it. Disposed in curves? The F. ray extends many minutes s, gradually fading away. Mar. 17, 1855. There seems to be a knot at p. extremity, in which the neb. terminates in that direction, and immediately s. of this knot is a little dark bay. The branch running f. from this curves round towards centre. See figure.
879	1	Apr. 16, 1852. F. brush; night bad.
881	1	Nucleus.
882	2	Mar. 22, 1857. mE. sp, nf. and bM.
887	2	Mar. 17, 1849. Dark space f. centre strongly suspected.
891 } 893 } 894 } 898 }	4	Mar. 26, 1856. Of this group 894 is the largest and brightest; its light is patchy.
895	2	Mar. 28, 1856. Irreg; R, edge ragged; sbM; nucleus.
896	2	Jan. 27, 1852. Neb. divided into two parts, and F. appendage np. Apr. 15, 1852. Black line across; comp. scarcely visible.
897	3	Feb. 22, 1857. lE. sp, nf; gbM. to F. nucleus.
901	1	Mar. 23, 1857. F; E. np, sf; lbM.
903	1	E; vbM.
908 } 911 }	3 {	Jan. 27, 1852. 908 mottled, with S. * involved; sp. it is a coarse d. *. 911 is irreg, with B. * in s. edge, and having dark lanes through it.
910	13	Mar. 30, 1856. Examined attentively for a long time; it appears to be of the shape annexed, which exaggerates; there can be no doubt of the bend upwards at α, and of the darkness about the nucleus; S. * at β. Apr. 6, 1856. Seen pretty much as before; the upward bend at α is at a right angle. The p. branch reaches as far as γ; and I suspect a S. * there. Mar. 8, 1858. This night is not as good as some on which I observed this object last year, but I can confirm my previous observations as to its general shape.
918	1	Mar. 3, 1851. E; in the meridian vlbM. Another brush-like 20′ np.
923	1	Mar. 22, 1857. vS; R.
925	2	There is an appendage, perhaps an independent neb; r?
930	3	Between 4 *s, in the shape of a trapezium.
931 } 932 }	1 {	Feb. 24, 1852. 2 rays, forming an angle of about 100°; the s. one has a nucleus, and there is a knot at the n. extremity of the other.
933 } 939 } 940 }	3	2 "novæ" near, probably a 3rd. 933 and 940 are E, the others R.
936	2	Apr. 1, 1848. A tolerably B. neb. with a smaller one f.
943	5	Spiral. Apr. 18, 1851. BM; F. neby. all round of a mottled character, knot or appendage in p. part. Apr. 10, 1852. Spiral? gbM. Mar. 1, 1854. Spiral arrangement; sky milky.
945	1	Apr. 11, 1850. Several *s near it, but few others in neighbourhood.
946	3	Mar. 8, 1858. S; lE. and pF.
947 } 950 } 951 } 953 }	2	All are S, R, and lbM.
948	1	bM.
959	1	Mar. 3, 1851. S; lenticular.
960	1	Feb. 17, 1855. A large number of pB. nebs. knots; I counted 8, probably there are more.
967 } 968 } 969 }	1	Mar. 28, 1856. The p. one is E. p. and f; the others are R; bM.

Number in Herschel's Catalogue.	Number of times observed.	Description.
971	1	Mar. 29, 1856. Neat little ray np, sf; bM.
973	1	
978	1	Mar. 13, 1852. Oval; F. nucl; another F; S; 5' nf.
980	1	Jan. 27, 1852. gbM; R; S.
981	2	April 13, 1855. Dull nucleus; edge ragged.
982	4	April 15, 1852. Spiral probably; knot in s. edge, and a * outside p. edge; another S. neb. 3' sf, having * immediately n. of it. April 16, 1852. Spiral; last night's observation confirmed; the spiral branch seems to start from the s. edge and go round the f. and n. sides as far as the * p. April 19, 1857. A * np. and a * in s. edge. Seen thus:—The spiral branch is B. and easily distinguished at sp. edge (α); as it extends to f. edge it grows fainter, and I can trace it no further than β. The central neb. is vB, and has a B. nucleus. The S. neb. sf. is BM. and a lE. Apr. 20, 1857. Examined with 1-inch and ½-inch single lenses; last night's observation is correct.
983 } 984	1	S. * p. 983 about 1'.
985	1	Mar. 27, 1854. vF; r?
988	2	Feb. 24, 1852. lE. n. and s; bM.
992	2	Mar. 7, 1851. E; bM; nucleus.
994	1	5' long.
1002	3	Mar. 17, 1849. Suspect it to be a spiral; though twe saw at moments ring round nucleus. Apr. 21, 1851. Spiral of the faintest class; the M. is pB, but the branches vF; conjectured form thus Apr. 17, 1855, or thus
1005	3	Apr. 11, 1850. Fine neb, but very bad night.
1006	1	vg. vlbM.
1008	2	* in nf. side; vF; E; B. nucleus.
1009	2	Apr. 13, 1852. Oval; gbM.
1011	4	Mar. 3, 1851. Lenticular; mottled. Mar. 30, 1856. mE. sp, nf; B. nucleus; very much mottled; the larger half of neb. lies to s. side of nucleus. A B. streak running obliquely through the nucleus, and another B. patch to s. end. Apr. 6, 1856. I see two patches in s. end, also a *. Apr. 19, 1857. Sketched. See fig. 17, Plate XXVII.
1014 } 1015	1	Apr. 14, 1852. The s. one is E; the n. one has 2 *s involved.
1017	1	Mar. 27, 1854. Filamentous; r; * near centre.
1018	3	Jan. 10, 1856. pB. nucleus in a L. mottled disc of F. neby. Irregularly R. Another nf.
1022	1	Jan. 27, 1852. Long ray; gbM.
1030	2	Apr. 15, 1852. The neby. p. centre is mottled.
1029 } 1031	1	The p. one is S. and the f. one vB.
1033	4	3 "novæ"; one is S. and R, the others are E.
1038	4	Mar. 27, 1856. mE. n. and s; smbM. to a B. nucleus; a d. * involved in n. extremity; a B. * further distant n.
1040	1	Apr. 13, 1852. mE. sp, nf; * p. a S. R. neb. about 7' np. it.
1041	1	Mar. 17, 1849. Roughly sketched; E, with a split or opening in the direction of major axis, and a * a little f. centre.
1043	5	Mar. 30, 1854. F; spiral? another neb. np. or nearly n; vF. about 5' distant. Apr. 6, 1855. Query, of this form? s N Its light is certainly patchy, and the neb. is lE. nearly p. and f; np. this object is another F. R. neb. with stellar centre. Apr. 13, 1855. Suspected shape as before, stellar centre. Apr. 16, 1855. My previous conjecture as to shape is rather confirmed by Mr. J. Stoney, who saw the p. branch turned off sharply to s. (nearly at a right angle), whereas the f. bend is not so sharp; but this latter branch reaches further round and is rather fainter. The whole object is vF. Mar. 27, 1856. Last year's observations fully confirmed.
1045	1	Apr. 26, 1851. Bicentral appearance is very indistinct; the light is mottled; E. ssp. and nnf.

M

Number in Herschel's Catalogue.	Number of times observed.	Description.
1048	2	Mar. 29, 1856. pL; B; mbM; r.
1049	1	Mar. 15, 1855. pF; R; lbM, but no nucleus.
1051	1	Apr. 18, 1851. S. ✻ involved in f. part of it, precedes a ✻ of 9th mag. 5′.
1052 } 1053 }	3	Jan. 27, 1852. Spiral. Apr. 9, 1852. Previous observations confirmed; S. ✻ np. it. Apr. 14, 1852. Drawing made. See fig. 18, Plate XXVII.
1058	1	Apr. 10, 1852. glbM; F. neby. round it; S. ✻ south.
1061	4	Apr. 27, 1851. Spiral; I suspect the f. branch extends to α. ✻ suspected at ρ.

Pos. Dist.
Nα 64° 2′ 50″
Nβ 256 2 19
Nγ 228 3 17
Nδ 15 3 53

Apr. 29, 1851. Observed for drawing. May 3, 1851. Viewed in twilight; drawn. Apr. 19, 1857. The p. branch seems the brighter rather of the two, and more suddenly curved than the f. one, and both of them look not quite so sharp as given in the drawing. See fig. 19, Plate XXVII. |
1062 } 1063 } 1064 }	1	Badly seen.
1066	2	Mar. 12, 1850. Broad equable band; several conspicuous ✻s in it, especially near ends.
1081	1	Apr. 16, 1852. vS. ✻ p. and a little n. of centre; I suspect another in n. branch; gbM.
1084	1	Apr. 16, 1852. Nucleus.
1085	5	Apr. 13, 1852. Brightest part a little eccentric; ✻ p. is involved. I suspect (Mr. B. STONEY) a dark curved passage on s. of centre, probably new spiral. Mar. 30, 1856. I have little doubt this is a spiral, either ✻s , which I rather believe, or , a S. ✻ p. Apr. 6, 1856. I think spiral with one branch; a B. part at α, and I suspect a ✻ there. Mar. 24, 1857. Nothing to add to previous observations, which, however, I can fully confirm. Apr. 19, 1857. Observed.
1088 } 1091 }	2	Apr. 21, 1851. 1st vF; 6′ ssp 2nd; 2nd vB. and mE; a d. ✻ 5′ nf, whose smaller component is blue.
1092	3	Apr. 1855. Two neb. about 14′ distant, 45° nf. Is the s. one of this shape, with a wedge-shaped division running downwards? The other neb. is lE. np. by sf; has nucleus, and is the larger and brighter of the two. Mar. 29, 1856. Last observation confirmed as to the shape of the s. one; the north one is, I think, a spiral of this shape; the branches vF.
1094	2	Feb. 26, 1851. A long ray, much resembling 242.
1105	4	April 13, 1855. pB; R; bM, but no nucleus.
1106	2	April 1, 1848. A very close cl. of faintish ✻s, preceded by a S. neb.
1107	2	March 9, 1850. A long ray with mottled light.
1108	3	April 17, 1855. Has a B. R. nucleus, surrounded by much F. neby, which is patchy and involves a B. ✻.
1110	1	April 13, 1850. E. np, sf; 88″ by 50″.
1111 } 1113 }	4	April 26, 1851. 1111 has a B. R. centre, with nucleus; then two dark spaces concentric, with nucleus; and outside these F. neby, as in figure. (δ) 1113 has F. nucleus, or stellar point.

αN 175° 2′ 25″
αβ 183 0 58
αγ 204 3 01
αδ 66 3 47

April 28, 1851. Previous observation rather confirmed; the dark spaces certainly exist, but I cannot be sure that appendages are not part of spiral branches. April 15, 1852. Last year's observation confirmed as to dark curved spaces p. and f. centre, and F. neby. outside them again. See fig. 20, Plate XXVII. |

Number in Herschel's Catalogue.	Number of times observed.	Description.
1117	3	April 25, 1854. R; has nucleus; * involved f. nucleus.
1119	1	Feb. 17, 1855. vB; R; bM; has 2 S. *s p.
1120 1121 1122 1124	1	Jan. 28, 1849. Observed in haze.
1128	1	Apr. 6, 1856. F; bM; a B. * in sf. edge and a patch in np. end. Neb. is fully 4′ long.
1129 1136	2	Feb. 26, 1851. The larger is vlbM; perhaps not R; S. one r.
1131	3	March 17, 1855. L. R; nucleus; * in nf. edge; mottled.
1132	1	March 9, 1850. A ray; diminution of light in neighbourhood of nucleus; edges parallel; night bad; remarkable object.
1140	3	April 6, 1855. Very like a distant cl.
1144	1	April 14, 1852. The brightest part in advance of the centre; vS. * n.
1146	6	March 8, 1856. Irreg. shaped neb. with nucleus eccentric, and a knot or appendage at f. end. March 27, 1856. There are 4 knots or *s in the neb, besides the B. patch to sf. side of nucleus.
1147	1	March 6, 1851. S. lenticular ray; B. nucleus.
1148	1	No description.
1149	1	March 15, 1849. Lenticular, with split in direction of major axis.
1155	1	No description.
1156 1158	1	April 6, 1855. Both are R; pB; bM.
1160	1	April 18, 1855. pF; L.
1162	2	April 25, 1854. E. p. and f; bM.
1167	1	March 9, 1850. Great ray; night bad.
1168	2	April 10, 1852. Has E. appearance np, sf; F. neby. all round it.
1171	1	March 13, 1852. E. p. and f; nucleus.
1173	7	See the 'Transactions' for 1850.
1175	4	April 20, 1857. A vL. B. E. neb; much mottled. The f. edges are comparatively sharp and well defined, but in the p. and n. edge there is a great inequality of light; nucleus E; vB. part to n. of nucleus.
1176 1180	1	April 16, 1852. gvbM; the f. one is much fainter.
1178 1183 1187 1189 1190 1194 1201	1	Apr. 13, 1852. The three or four brightest are E; gbM.
1179	1	No description.
1185	2	April 10, 1852. L; E. p. and f; arms F; * involved in p, and another * in f. arm, but a little further from centre.
1186 1188	2	April 26, 1849. 3 in line; the f. one vF, the other two R; pB. nuclei.
1195	1	April 10, 1852. F. knot at end of p. branch.
1197 1200	1	{ April 14, 1852. 1st E. * in np. extremity; 2nd F, almost planetary; another vF. and thin ray about 30′ f.
1196 1202	5	{ Sketched 4 times. March 1, 1851. 1196 is bM, and has a vF. comp.; 1202 is a spiral, B. centre, and 2 knots. There is another neb. 10′ nf. About 84° 34′ N.P.D., } There is a scarlet * 10m. and a F. E. neb. 10′ s. of it, with *s and 12ʰ 25ᵐ Æ } in it. See fig. 21, Plate XXVII. April 9, 1852. Last year's observations confirmed.
1204	2	March 15, 1855. pL; mbM. to a sharp nucleus; mE. p. and f.
1209	1	April 24, 1854. Lenticular nnf. by ssp; F; vlbM (night bad).
1211	1	March 9, 1850. Spiral; a F. neb. f; roughly sketched.
1212 1221	1	{ Feb. 17, 1855. 1212 is B, R, and smbM. 1221 is vB; mE. sp, nf, with a suddenly B. centre.
1225	3	April 13, 1855. vB. globular centre; E. p. and f.

Number in Herschel's Catalogue.	Number of times observed.	Description.
1231	1	April 10, 1852. Not R.
1232 1236	1	March 6, 1851. vbM; edges fade off. 1236 is vF.
1237 1250	3	March 13, 1852. 6 knots, one E. March 1, 1854. One has dark spaces about the nucleus. March 15, 1850. 12 knots examined.
1239	1	No description.
1240	1	April 24, 1854. R; B. nucleus; outline somewhat irregular.
1242 1251	1	March 6, 1851. Both BM; B. * involved in 1st; 2nd is E.
1245	4	March 30, 1856. pB; E; nucleus. A B. streak runs up through the nucleus, growing broader at p. end; on either side of this I suspect dark spaces, and outside them again F. neby, especially to s. side of nucleus. April 19, 1857. Much E; instead of a nucleus it has a vB. narrow central streak; to the left of this I suspect a darkness; then outside this more F. neby, as in sketch. See fig. 22, Plate XXVII.
1252	2	April 17, 1855. There are here 4 neb; the 3 f. ones seem to be involved in a mass of F. neby.
1253	1	April 15, 1852. gvmbM; oval. Another 14' sp; also vB.
1258	4	Sketched 3 times. April 12, 1849. Uncertain whether a d. nucleus, or nucleus and *; neb. decidedly darker in middle, following the nucleus, and rather brighter outside this. March 7, 1856. d. nucleus, or nucleus and *, which are eccentric, being nearer the sp. side; light uneven and patchy; suspect a darkness nf. the nucleus. March 8, 1856. Last night's observation confirmed. March 18, 1857. Seen as in the rough sketch subjoined; a * close sp. nucleus.
1262	1	Feb. 16, 1855. E. sp, nf.
1271	1	April 25, 1854. L; svmbM. to a nucleus; pmE. p. and f.
1274 1275	1	April 13, 1849. Found in this set 11 knots, of which 6 are 1203, 1237, 1244, 1253, 1274, and 1275; the remainder are "novæ," one of these latter being hollow in middle; probably a ring seen obliquely; a F. * n. of its middle; seen best with single lens. Remarkable object.
1280	2	March 26, 1856. E; B. nucleus; F. extremities.
1281	1	March 17, 1849. 3 nuclei, or 2 nuclei and *, and F. neb. outlying.
1282	1	April 11, 1852. gbM; oval; E. n. and s.
1286	1	April 18, 1855. Like a distant cl; vB. nucleus.
1294	1	Feb. 26, 1851. 4 found.
1296 1298 1301		April 10, 1852. 1301 is vgvbM. 1298 is smaller, and much the same character.
1306 1308	4	Sketched twice. March 28, 1856. A rough sketch made; suspect spirality in the n. one; the large neb. has an appendage n. of nucleus and a little f. it. March 24, 1857. Examined to confirm drawing, which I think is pretty accurate. See fig. 23, Plate XXVII.
1309 1315	2	A d. neb.
1312	2	March 9, 1850. Another spiral; dark spaces, especially one sf. nucleus.
1332	1	
1333	1	April 22, 1854. E. n. and s; nucleus vB; light uneven.
1337	2	April 19, 1855. Seen by myself, as represented (see fig. 24, Plate XXVIII.). Mr. STONEY, who was with me, did not see the F. curve at p. extremity, which therefore needs verification. I myself felt pretty certain of it. March 29, 1856. Seen as last year; sketched. See fig. 24, Plate XXVIII.
1343 1348	1	April 16, 1852. gvbM; 2 others, both E. about 20' s. of 1348.
1345	3	Feb. 19, 1855. E. p. and f? B. nucleus.
1352	2	April 11, 1852. A vL. ray; gbM; some *s involved.
1357	3	Roughly sketched twice. April 17, 1855. A beautiful object, very well seen in finding-eyepiece; the whole neb. (taking into account the appendage) is much broader at nucleus than elsewhere, narrowing off suddenly, and the nucleus projects forward into the dark space; and immediately opposite this the F.

Number in Herschel's Catalogue.	Number of times observed.	Description.
		appendage is broadest and brightest. The ray is 12′ or 14′ long, and there is a F. ✲ at α (Mr. Stoney was with me). April 6, 1856. 15′ long, perhaps even longer; the ✲ opposite the nucleus is about two-thirds the breadth of the neb. distant.
1358 } 1359 }	2	April 14, 1852. A curious d. neb; some other nebs. p.
1362	3	March 19, 1857. R; bM; L. but F; ✲ involved in p. edge.
1363		Looked for twice; not found. Query, Is this 1358 and 1359?
1368	3	May 3, 1851. gmbM; 1E. sp, nf; edges fade off very gradually.
1382	2	April 24, 1854. pB; has nucleus; E; F. ray f. April 25, 1854. Seen as last night, also the F. ray f; about 50′ p. is a B, R, pL. neb. f. a B. ✲.
1385 } 1392 }	6	Sketched 3 times. April 10, 1855. Somewhat curved, like 2205. The s. branch is patchy, having 2 B. spots (see fig. 25, Plate XXVIII.); the n. branch is much the brighter. A S. ✲ p. the neb. About 6′ or 7′ n. of 1385 and a little f. is 1392, not so F. as Herschel describes it; the brightest part seems eccentric, being nearer the nf. edge. From this B. part I suspect a curve round n. to sp. April 13, 1855. Seen as before. March 8, 1856. The comp. n. (1392) I suspect, as before, to be a F. spiral. March 27, 1856. Better seen than on any previous occasion; the F. branch to the left extends round as far as the p. extremity of the B. branch. The comp. neb. suspected to have a twist in it, as before; sketched.
1386	2	April 11, 1852. gvbM; E. np, sf.
1397		See the 'Transactions' for 1850.
1402	1	March 1, 1854. d. neb; F. neby. connects them.
1403	2	April 22, 1854. A remarkable object; spiral?
1408	2	April 5, 1851. L; vB; comp. neb. pB.
1409	1	March 7, 1856. 1E. nearly p. and f.
1411	1	Feb. 26, 1851. p. part is broadened out; light unequal; night bad.
1414 } 1415 }	3	April 26, 1851. Herschel's two neb. form one, the joining part in middle F, and vF. production of neb, as in sketch. April 9, 1852. Last year's observation confirmed; like a caterpillar on a leaf. April 20, 1857. I can confirm former observations in every particular, and think there are two additional ✲s in f. part. See fig. 26, [Plate XXVIII.
1431	3	Spiral?
1436	1	April 13, 1852. gvmbM; S. ✲ involved in f. part.
1437	1	April 15, 1852. gvmbM; oval.
1441	8	Feb. 16, 1855. vB. ray; a dark band across on each side of nucleus, separating it from the extremities. Feb. 17, 1855. Sketched. Feb. 19, 1855. The dark spaces which are visible in finder are not black, but only portions of fainter neby. April 6, 1855. Seen as before; dark lines very plain in finder. April 16, 1855. My sketch exaggerates the dark lines; they should be broader, and not so well defined. Mr. Stoney remarked a second dark line across the n. branch near its extremity. Mar. 7, 1856. Observed. Mar. 18, 1857. Dark spaces far apart, and not absolutely dark; suspect a dark space to right-hand side of nucleus. See fig. 27, Plate XXVIII.
1451	4	March 9, 1850. Another spiral; another neb. 15′ p. Feb. 26, 1851. Spiral; 2 arms, and some ✲s in f. arm; centre is B. 12′ p. and a little s. is another neb, E; and 30′ nf. is a 3rd, E. n. and s. April 15, 1858. vL. and vB. The centre itself is like an E. neb, with nucleus; this centre is enveloped in an irreg. ring or rings of nebulous light, as in the accompanying rude sketch, which does not contain all the details. sp. this object there is a S. neb. E. np, sf. and very patchy, and I suspect it to have a F. nucleus. May 3, 1858. I saw all the details in last observation, except that there was only one ✲ visible s. of nucleus instead of two, but this is not quite so good a night. The surrounding ring of neby. is of irreg. shape; it curves gently at δ, but bends more sharply at γ, where it is brightest. The centre seems to reach up to and blend with the neby. at δ.
1456	5	April 9, 1852. Spiral; bears great resemblance to 1111. April 14, 1852. F. neby; 2′ radius extends all round, in which I think I see traces of spirality which exist certainly in the central part (Note by Mr. B. Stoney). A good night and speculum in good order would probably show this object distinctly. April 13, 1855. vlE. p. and f; dark ring round the nucleus; then B. ring exterior to this. The annulus, however, is not perfect, but broken up and patchy, and the object will probably turn out to be a spiral. There is much F. outlying neby. March 8, 1856. Annular at first look, but ring not perfect; centre vB.

Number in Herschel's Catalogue.	Number of times observed.	Description.
1460	1	April 25, 1854. mE.
1462	3	March 7, 1851. gvmbM.
1466	2	March 7, 1851. 8′ long; R; centre vB.
1475	3	March 1, 1851. Nucleus 2′ s. of * of 10th mag. At Æ 12ʰ 43ᵐ and N.P.D. 60° 20′; "nova," with nucleus; E.
1486	11	March 11, 1848. Curious circular-shaped neb, with a dark and large spot at one side, around which is a close cl. of well-defined little *s. May 4, 1851. E. nearly p. and f. Herschel's dark space is a curved passage, extending from p. round the f. side of the nucleus by the n.
1498	3	Feb. 16, 1855. L. B. ray; nucleus oval and vB; there is a * involved in n. edge, a little preceding the nucleus.
1499	1	Apr. 17, 1855. vF; mE. sp. nf; has a plainly seen * at n. end, and either a * or what looks more like a B. little knot involved in s. end.
1500	1	Numerous neb. around.
1509	4	Apr. 18, 1855. Looks sometimes like 838 when badly seen, with a B. E. patch in centre and dark spots on each side of this; sometimes dark ring is seen all the way round, but blackest to right and left. The neby. round it is mottled. In looking for this I found at about Æ. 12ʰ 48½ᵐ, and a little n. of this set a F. d. neb. E. at right angles to each other. Mar. 29, 1856. Last year's observation correct; * in sf. edge; sketched. Apr. 24, 1857. Long and carefully examined; the B. centre is E. in the direction of * on edge, and on either side of centre there certainly exist dark spaces, as before remarked, giving it the look of 838; yet sometimes I thought I saw it with a break in the outer annulus.
1515	1	Mar. 24, 1857. vvF; lbM; vlE. np, sf.
1525	1	Apr. 27, 1854. vF; R.
1536	1	Apr. 18, 1855. Like a distant cl; 2 B. *s involved.
1547	3	Mar. 12, 1852. B. lenticular ray with E. centre. May 3, 1858. Sketched; like 2172 and 1357.
1549	3	May 3, 1856. gbM; B. nucleus; B. * in np. end. April 15, 1858. Very much mottled.
1551	1	Apr. 22, 1854. E. p. and f; bM.
1556	1	Spiral?
1558		cl.
1559	3	Mar. 12, 1852. Light equable; E. sp, nf.
1562	1	Mar. 24, 1857. vF; lbM; lE.
1564	2	Mar. 1, 1851. vB. centre; has an appendage parallel to major axis.
1569		cl.
1570	3	Spiral? darkness sf. nucleus.
1576 ⎫ 1577 ⎬ 1578 ⎭	3	A group of 4.
1580	2	Apr. 25, 1854. R; bM; between 2 *s.
1589	7	Sketched three times. Apr. 29, 1856. The B. centre is E. but not in the direction of the neb. The whole neb. is much mottled. Apr. 15, 1858. I can add nothing to my drawing and observations of last year, which are fully confirmed. See fig. 28, [Plate XXVIII.
1599 ⎫ 1600 ⎭	2	{ Apr. 13, 1855. Both are S; R; pB; bM. Apr. 17, 1855. There is a 3rd vF. neb. nearly n. of the f. one of these two.
1604 ⎫ 1605 ⎭	2	{ Mar. 28, 1856. 1604 is lE; pB; has nucleus; and a * at np. end; 1605 is R; vF; and its light equable.
1622	19	Carefully observed since drawing published in the 'Transactions' for 1850. The outer nucleus unquestionably spiral, with a twist to the left; thus
1626	2	Apr. 19, 1855. Oval; bM; * np. May 3, 1856. About 5′ nf. it is a vF. nebs. knot.
1638 ⎫ 1639 ⎬ 1643 ⎭	2	{ May 3, 1856. 1638 is E; nearly p. and f; bM; 1639 is S; R; bM; 1643 is the largest; pB; R. and gbM; nucleus, round which I suspect dark spaces.
1647	1	Apr. 19, 1855. Not L; gbM. to a nucleus; mottled.
1650	3	Apr. 19, 1855. L; pb; B. nucleus; seen as in sketch, but not certain whether the lower branch joins the nucleus, or is only the continuation of the upper curve. Mar. 21, 1856. The p. arm *does* appear to originate from the nucleus, which is vB. and oval-shaped. Mar. 30, 1856. Seen as before. See fig. 29, Plate XXVIII.
1658	3	Mar. 28, 1856. F. ray n. and s; no nucleus; light; equable.
1659	2	Mar. 24, 1857. S; vF; nearly R; brightest part is on sp. side of centre.

Number in Herschel's Catalogue.	Number of times observed.	Description.
1663		Splendid cl.
1664	3	Mar. 27, 1856. pL; pB; R; sbM; about 2' or 3' f. is a S. F. neb.
1668	1	May 3, 1850. A single B. * at n, and a d. * at s. end of this neb. Another neb; R; bM; sp.
1669	1	Apr. 11, 1852. A vF. amorphous-looking neb; S. * in s. edge.
1672	1	Mar. 1, 1851. * or nucleus in np. edge; 2nd vF; 3' s; both E. p. and f.
1676 1679 1680	2	All F.
1695	1	May 15, 1854. vF. in twilight; lbM.
1697	2	Feb. 19, 1855. mE. p. and f; L; pB; gbM. Mar. 24, 1857. Found here 3 neb. in a line sp, nf; all of them are bM.
1703	1	Apr. 14, 1852. gbM; L. vF. neb. 14' s. of 1703; also a S, F, E. neb. 15' p. and 2' n. of 1703.
1711	4	Mar. 28, 1856. S; bM; dull nucleus; lE.
1713	6	Apr. 24, 1854. Centre pB; oval n. and s, and among several *s; I thought the n. end the broader, and suspected a dark space p. the nucleus. May 1, 1854. Singular object; the main body of neb. has a B. nucleus, and is E. n. and s; the southern end bends back suddenly at a sharp angle, and extends np. past the neb, ending in a B. R. patch or nucleus; 3 *s around the neb. Apr. 17, 1855. Mr. STONEY saw the p. branch extend *round* the s. end of the main neb. and continue on to n, when after a second turn it joined the nucleus. See fig. 30, Plate XXVIII.
1714	3	Feb. 19, 1855. pB; R; bM. to a nucleus.
1715 1716	1	Mar. 9, 1851. 3 found; all S; F; R.
1734 1735	5	Apr. 18, 1855. The n. one is spiral?; 3 *s in it; to myself it appeared to have a single branch running from below the nucleus round the n. and f. edges. Mr. STONEY suspects two branches. May 10, 1855. n. one suspected spiral as before; the s. one is, I think, lE. n. and s, and the * between the two neb. is d.? Mar. 29, 1856. Suspect n. one as before; it is a very difficult object, and requires a fine night. Apr. 24, 1857. Last observation fully confirmed as to spirality of the n. one. I still think it has but one branch. The * between the 2 neb. is d.
1741 1742	3	Mar. 28, 1856. 1741 is S; R; bM; pB; 1742 is S. ray nf, sp, and has a * of 12th mag. at its s. extremity.
1743	1	Mar. 17, 1855. mE. p. and f; nucleus. Query, a knot in p. branch.
1744	8	Sketched 3 times. Mar. 1, 1851. Large spiral; faintish; several arms and knots; 14' across at least. See fig. 35, Plate XXIX.
		April 27, 1851.
		αN 195° 1' 22″
		αβ 345 1 50
		αγ 273 3 31
		αδ 74 3 1
		αε 74 1 38
		αn₁ 99 5 37
		αn₂ 135 5 2
		αζ 10 5 19
		αη 358 5 33
		αθ 44 5 1
		αι 240 8 34
		May 3, 1851. ακ 118 3 34
		αλ 112 3 29
		αμ 72 4 0
		αν 211 6 31
		γℓ 263 2 56
		γσ 220 3 47
1745	1	Mar. 29, 1856. Has a nucleus; light very patchy; 3 *s in edge; vF. Query, spiral, with a right-handed twist? About 4' f. is a S. pB. E. knot.

Number in Herschel's Catalogue.	Number of times observed.	Description.
1746		Close, rather F. cl.
1754	2	Mar. 27, 1856. S; bM; mE. np, sf.
1755	1	Mar. 9, 1851. E.
1757	1	Mar. 29, 1856. 2 neb. 3′ apart; n. one vS; bM; the other a ray p. and f; nucleus.
1762	2	Apr. 13, 1850. 3 knots near.
1764	1	Apr. 19, 1855. Long narrow ray, with a S, R, vF. neb. sf. About 15′ np. of 1764 is another vF; and about 6′ p. and 1′ n. of this last is another eeF.
1766	2	Apr. 13, 1852. bM; S. ✳ s. of it. Mar. 30, 1856. E. nearly n. and s; S. ✳ sf; B. nucleus.
1768 1769	1	Mar. 1, 1851. 1768 S; F; E; 1769, nucleus.
1770	1	Mar. 6, 1851. Another 5′ p, and another 10′ sp; vF.
1771	5	Apr. 10, 1852. Either a d. neb, or 2 knots of one neb.
1773	2	Apr. 29, 1856. mE; not F; lbM; major axis sp, nf.
1774	1	May 10, 1858. S; irreg; R.
1776		Frequently observed; nothing certain.
1778 1779	4	"Nova" near; 1st E, 2nd bM, 3rd vF.
1782 1783	4	{ May 12, 1850. 1782 pB; L; gbM. 1783 vB; R; nucleus. Another L. F. ray about 16′ nf. 1783.
1788 1789 1791	2	Only two found; both S; R; bM.
1790	1	May 15, 1854. pL; vF; lbM.
1792 1793	2	Both F.
1797	2	Mar. 28, 1856. R; pB. Its brightest part is nearest f. edge, and forms a curve round n.
1799	2	Mar. 29, 1856. pL; lE. n. and s.
1804	2	Mar. 1, 1851. d; bM; two others F.
1805	1	Mar. 28, 1856. Long narrow ray; F; bM.
1813	2	Apr. 13, 1852. vgvlbM; filamentary appearance of the branches quite apparent. Though unmistakeably a cl, yet on a very bad night it would be seen as a neb.
1815	1	Apr. 17, 1855. E. sp. nf; nucleus.
1817	1	A B. d. neb.
1818	1	Apr. 19, 1849. r?
1820	2	Apr. 9, 1852. S; bM.
1825	1	Mar. 6, 1851. vlbM; S. ✳ f.
1829	1	Mar. 1, 1851. Nucleus; lE.
1833	1	Apr. 14, 1852. R; vlbM.
1835	1	Mar. 29, 1856. pL; gbM; ✳ in f. edge; between this ✳ and the centre the neb. seemed black.
1840	2	May 10, 1858. lE; vF. and flickering.
1842	2	Apr. 11, 1850. Narrow ray; bad night.
1843	1	Apr. 26, 1851. Within trapezium of 4 or 5 ✳s; lE. n. and s; vlbM.
1844 1845	1	{ Apr. 13, 1855. The p. one is lE. p. and f, and is the larger of the two. The other is S; R; pF. and bM.
1848	2	Apr. 11, 1850. E. Central part seems unsymmetrically placed with respect to general fig. of neb. Apr. 13, 1850. 3 "novæ" near; one of them mottled, and ✳ in s. border.
1851	3	Apr. 16, 1858. S; B; with B. sharp nucleus, and ✳ involved n. of nucleus; 2 "novæ" near.
1861 1864 1865	1	{ Apr. 16, 1855. 1861 is a narrow ray; 1864 is S. and R; 1865 is quadruple, and suspected to be one neb. connected by F. neby.
1854	6	Nucleus; dark ring suspected, like 450, but no conclusive evidence.
1857 1863	3	{ Apr. 7, 1851. Light mottled; another f. about 12′, and a little s; E; bM. Apr. 13, 1852. Spirality suspected.
1870	2	r; lbM.
1872	2	May 10, 1855. pB; R; nucleus; E; mottled.
1873	2	May 15, 1854. pB; S; R; bM.
1874	2	Apr. 25, 1848. E; ✳ at each end.
1879	2	May 14, 1855. vF; nucleus or ✳ in centre.
1880	2	May 16, 1855. 2 neb, with 3 B. ✳s in the neighbourhood. Both F. and E.

Number in Herschel's Catalogue.	Number of times observed.	Description.
1881	3	May 12, 1858. Close d. neb.
1883	1	May 14, 1851. pL; mE. ssp, nnf; has nucleus.
1885	3	May 12, 1858. Rather a B. ray; bM. and mottled. Its p. arm is brighter than the f. one. A F. neb. about 2' p.
1890	3	May 1, 1854. F; pL; no nucleus; mottled.
1891 1893 1895	2	Apr. 3, 1854. 3 neb. in line, and another S. neb. near the f. one.
1892	3	May 1, 1854. Has a curved form between 2 *s, and in contact with them; there is a 3rd smaller * close to the neb. on np. side.
1894	4	Mar. 1, 1851. B. in centre; E.
1898	2	Apr. 6, 1851. E. p. and f; r. Another vF; 3' f.
1901	1	Apr. 19, 1849. 6 neb. found.
1903	1	May 3, 1851. E.
1904	1	Mar. 17, 1855. The atmosphere seems to exist.
1905	6	Apr. 28, 1848. Think the distance between the 2 neb. greater than in H's drawing. Apr. 11, 1850. The 2 neb. not in a line, and a F. connexion suspected. Apr. 17, 1855. These 2 neb. are not in a line but parallel; the distance between is considerable, but F. neby. suspected connecting them; they have a very hazy look, and the edges are not well defined. May 14, 1855. Seen as on last time. May 8, 1861. Sketched; axis not parallel, but inclined at an angle of about 16°. Fig. 31, [Plate XXVIII.
1907	3	May 3, 1856. A B. S. ray sp, nf; has nucleus.
1908	2	May 16, 1855. Looks R. pB; mbM; nucleus.
1909	8	Apr. 13, 1850. vB; oval; E. np, sf; * in np. end. "Nova" near; vS.
1910	2	Apr. 13, 1855. mE. n. and s; centre vB; extremities F.
1911 1912	2	Apr. 25, 1849. 1911 vS; * close to right. 1912 rather F; vS. * involved. "Nova" f. and vF.
1913	1	2 "novæ" f, apparently connected.
1914	2	Nucleus, and E.
1915	3	May 23, 1854. 2 neb. close together, n. and s.
1916		A superb cl.
1917	5	Apr. 13, 1850. Very remarkable ray, 12' or 15' long; α, β, γ, and δ are *s, of which α is F; a long split precedes the nucleus.
1919	2	Spiral? About 15ʰ 12ᵐ Æ } A pair of new neb, about 15' asunder, np. and sf. The sf. one a pB. ray, 33° 55' N.P.D. } the other F. and S, but neatly placed at one angle of a triangle of F. *s.
1920	2	Apr. 28, 1851. vlbM; E. p. and f.
1923	4	Mar. 17, 1855. S; R; bM.
1924 1925	1	Apr. 11, 1850. Elegant little d. neb.
1926	2	May 3, 1856. pF; R; bM.
1927	1	Apr. 13, 1852. 3 neb; 2 pB, the 3rd S; E; F; 15' sp. 1927.
1928	2	Mar. 17, 1855. E. np, sf; bM; not vF.
1929	3	F. dash of light.
1930	1	Apr. 3, 1854. R; B. nucleus.
1931	2	Apr. 13, 1855. Has a ragged edge and mottled look; about 6' or 7' nf. there is another.
1934	6	May 6, 1850.

May 6, 1850.

	Pos.	Dist.
AB	288°	7' 53"
BC	299	6 28
CD	283	8 23

Suspect (A) to be a spiral, to be re-examined on a fine night. (B) a B. condensed oval neb. (C) vF. ray. (D) eeF; S. neb. May 14, 1850. (A) Dark spaces round on either side of nucleus, seen at moments; also a dark line running along the sf. edge, splitting off a part of neb, which has a B. knot to s, also some ill-defined dark space at n. end. Apr. 5, 1851. (A) spiral; a good deal of dark space round the nucleus, branches perhaps like 604.

Number in Herschel's Catalogue.	Number of times observed.	Description.
1936 1937 }	2	May 23, 1854. d. neb; both pB; R, and mbM.
1938	3	Mar. 9, 1851. BM; S. neb. p.
1939	1	May 30, 1851. Lenticular ray; bM; 4 *s close s.
1942 1943 }	4	May 14, 1855. Both S; R; lbM; S. * closely nf. the s. one.

1946	11	May 5, 1850. Strongly suspected to be annular neb. with * near the centre. Apr. 5, 1851. Like 450; dark ring plainer seen on p. part of neb; very S. * n; about ¾ diameter of neb. off. The f. part of dark ring a little broader than the p. part. May 3, 1851. Distance between nucleus and S. *
		
		May 3, 1851. Dist.
		                0' 26"
		                0 32
		                0 31
		            Pos. Dist.
		May 4, 1851. 7° 0' 25"
		              5 0 28
		              7 0 28
		
		May 29, 1851. The S. * scarcely seen; dark ring not at all. Apr. 3, 1854. The dark ring round nucleus seen pretty well; also the minute * n. of neb. See fig. 32, Plate XXVIII.

1947	2	May 22, 1854. vL; F; oval.
1950	1	vvF.
1952	1	vF; vlbM.
1953	1	Apr. 7, 1851. eF; 2 or 3 *s in edge.
1958	1	May 26, 1849. 2 new neb; one eF, the other S; 1958 R; bM.
1960 1962 1963 }	2	Another near.
1964	1	Apr. 19, 1855. S; F; R; bM. Another neb. 4' nf.
1968		cl. in Hercules. May 6, 1850. Seems to have a dark streak across the B. part a little above the centre. Apr. 6, 1851. Dark lanes seen which bear some resemblance to those in Neb. Andr. Apr. 27, 1851. Sketch made; dark spaces seen through mist. May 3, 1851. Sketched. May 26, 1851. Sketched. Apr. 17, 1855. The dark lanes are quite discernible in the finder eyepiece; they do not meet in the centre of the cl, but to sff. of it (see fig. 33, Plate XXVIII.).
1969	2	Apr. 19, 1855. The nucleus is nearest the p. edge, and light mottled.
1970	8	May 5, 1850. Intense blue centre fading off to some distance all around; S. *s to nf, to which neb. nearly extends. May 12, 1850. I fancied once or twice there were projections p. and f. (N.B. The existence of these not satisfactorily proved.)
1971		cl; in finder eyepiece the branches have a slight spiral appearance.
1972		cl; May 30, 1851. A dark lane above the centre quite across, or rather the upper one-sixth of cluster is much fainter than the rest.
1979		cl; June 3, 1851. The outline not R; on s. side is an outlying portion separated from the chief portion by a dark passage.
1981	2	May 31, 1851. I suspect annular, but twilight leaves me quite uncertain; n. edge is the brightest. June 3, 1851. Annular, n. edge is the brightest.
1983		cl; *s S. and very close together.
1989	1	May 29, 1851. Seen in twilight; looked very like a * of 9th mag.
2019		cl.
2023	11	Never well seen on account of twilight. Nothing additional since 1844, except a pB. * sf. middle.
2036		cl.
2037	5	Aug. 28, 1850. Annular or perhaps spiral, and * distinctly seen in dark part. The dark space is undoubtedly irregular in its form. Aug. 24, 1851. Annular; centre very suddenly darker than the rest of the neb; vS. * in np. edge of central part.
2042		cl.
2043	2	Aug. 1, 1851. 4 *s in neb, and 2 more on p. edge.
2045	3	On very bad nights.
2046		cl.
2047	7	Aug. 31, 1850. Centre rather dark. Aug. 1, 1851. The dark part is a little np. middle.

Number in Herschel's Catalogue.	Number of times observed.	Description.
2049		cl.
2050	6	Aug. 28, 1850. A very remarkable object, perhaps analogous to H. 450. The ring is not easily seen, but there can be no mistake about it; under the central ✳ there is a darkness. Aug. 22, 1851. sE. np, sf.
2060	13	Places of principal stars laid down, and a new drawing made. First observation Aug. 10, 1850, and last, Aug. 30, 1851. See end of Catalogue, and fig. 43, Plate XXXI.
2064		cl.
2071		cl.
2072	8	Aug. 23, 1851. Fine annular neb. like that in Lyra; R; the dark space is sE. p. and f; ✳ easily seen in np. edge, others suspected. Aug. 19, 1855. There is a conspicuous ✳ on the inner edge of the ring at np. side, and another fainter near this on the outer side. I believe the whole of this corner of the annulus is r, and can see the ✳s sparkling near the two already described.
2075	11	Roughly sketched five times. Aug. 10, 1850. ✳ or B. nucleus nf. the middle. A dark curved line p. this plainly seen, which at moments I fancied went round the sf. part. Sept. 9, 1852. This planetary neb. is a beautiful little spiral. Aug. 12, 1855. I think spiral, of the shape annexed. Aug. 16, 1855. The night bears ½-inch single lens *well*. There is a group of 4 minute ✳s p. the neb. Sept. 6, 1856. The details in my sketch of last year seem correct. I can trace the spirality distinctly. See fig. 34, Plate XXVIII.
2079 2080 }	1	2079 vF; E; 2080 vF; S; R.
2081		cl.
2084	8	Sept. 6, 1850. New spiral, with three branches, of which two terminate in knots, as in sketch; a fourth branch suspected. Sept. 8, 1850. Examined and drawing made.

Sept. 9, 1850.

	Pos.	Dist.
cμ	87°	4′ 10″
cl	263	2 47
cα	158	1 41
cθ	67	3 34
cγ	221	2 21

Aug. 21, 1851.

ac	325	1 36
αβ	213	0 33

Aug. 23, 1851.

αγ	257	2 08
αδ	261	2 27
αζ	302	2 03
αl	281	3 26
αθ	37	3 46
αε	47	1 04
αμ	67	3 46
αι	162	1 56
αϰ	169	2 14

F. branch (D) p. centre seen. Sept. 6, 1855. The two f. branches A and B unite in one before meeting the nucleus. I certainly see a fourth branch D, which seems to join C in the same way before reaching the nucleus. Of the four, those which terminate in knots are the brightest. B is fainter, and D much fainter still. See fig. 36, Plate XXX.

Number in Herschel's Catalogue.	Number of times observed.	Description.
2086	1	Aug. 21, 1857. R; vS; lbM.
2087 2089 }	2	Aug. 27, 1857. A group of 5 neb; many ✳s among them.
2088	10	Aug. 5, 1851. The nebula resembles the Milky Way, and is full of dark uneven rifts or lanes. The p. edge is the brightest, and the M. is darker than the edges. Sept. 6, 1856. There are portions of its p. edge clearly r.
2090		cl.
2092	3	Aug. 5, 1851. Resembles the neb. 2088, though on a much larger scale; the dark spaces have a rounder or more sack-like appearance, especially at the chief bend, where the neb. is also the brightest. It has several outlying portions of flocculent neby, especially at s. end. Sept. 3, 1855. General shape that of Herschel's figure, but several dark bays in it, and many more ✳s seen in and about it.

Number in Herschel's Catalogue.	Number of times observed.	Description.
2095	1	Aug. 27, 1857. eF; vlbM; no nucleus; E. n. and s.
2097	1	Sept. 5, 1850. R; bM.
2098	11	Since published in the 'Transactions' for 1850.
2099	6	Aug. 19, 1855. The neb. has 3 knots in it; a drawing taken. Sept. 3, 1855. Seen as before, and sketch compared. Sept. 6, 1855. Observed. Sept. 6, 1856. Details as in sketch confirmed. It is vB. See fig. 37, Plate XXX.
2102	3	Sept. 29, 1850. 2′ long; E; nucleus.
2106	1	
2109	1	Aug. 27, 1857. vvF; irreg. R.
2110		cl.
2112	3	Sept. 3, 1856. bM; edges indistinct; a * in nf. edge.
2120		cl.
2121	2	vF; lE. nearly n. and s.
2122		No definite cl, but sky thickly studded with stars.
2125		cl.
2127		Loose cl.
2128		cl.
2130		A red * of about 12th mag. in a scattered cl.
2132	6	Sept. 18, 1857. Centre r; mottled; * in edge.
2133		Searched for four times; not found.
2135	2	vF.
2139	11	Form not distinctly made out. See fig. 38, Plate XXX.
2142	1	Not well seen.
2143	5	Never well seen, being very low.
2146	2	E; lbM.
2149	14	Sept. 16, 1854. There can hardly be a doubt that this neb. is a cl.
2150	4	Oct. 23, 1857. lE. sp, nf.
2151	2	Sept. 20, 1857. There is a twist in the neb, but it is so F. that I cannot make out its shape.
2152	1	vF; lbM.
2154		A poor loose cl, with red * of 9th mag.
2156 } 2158 }	1	First has * in nf. edge, and is bM; the other is R; no nucleus.
2157		cl.
2160	4	S; nucleus; forms a quadrilateral with 3 *s; F. outlying neby. extensive.
2162	1	bM; E. p. and f.
2163		cl.
2164	2	Oct. 23, 1857. A vF. ray.
2165	6	About 24′ p. and 10′ n. is another vF; E. np, sf; 80″ long, 10″ broad. 2165 has a sharp nucleus, and is S.
2166	2	vF; pL.
2167	5	Oct. 2, 1856. * in centre; mottled; and * or knot in sp. edge.
2168	2	Oct. 7, 1855. E. n. and s; a vF. * nf. centre; centre B; extremities vF.
2172	17	The sketch conveys accurately the results of these observations. There are 5 knots near. See fig. 39, Plate XXX.

Sept. 12, 1849.

AB	62°	63°
AC	51	
BC	23	21
BD	116	
AD	96	98
BE	163	
AE	119	123
DE	243	
Direction of A	174	

Number in Herschel's Catalogue.	Number of times observed.	Description.

2173 } 2175	9	Oct. 7, 1850. Upper neb. is equable in light, and is much the fainter.
		Sept. 1849. Position of B 91°
		                Position of A 159
		                                       Dist.
		Oct. 7, 1850. AB 97° 5' 23"
		                A 157
		                B 91

2176	2	Sept. 20, 1857. Narrow ray sp, nf; vvF.
2178	1	Planetary?
2179	2	S; R; bM; nucleus.
2180	1	vvF.
2181	1	F; S. 7 knots found.

2183 } 2184	4	Pos. Dist.
		Nov. 27, 1850. αβ 235° 5' 29"
		                  αδ 73 5 44
		                  αγ 18 5 41

2185	3	Oct. 7, 1855. S; R; pB; mbM.
2186	3	Aug. 30, 1851. E. np, sf; light uneven.
2189	4	A group of 4 neb.
2191	1	No description.
2195	3	A group of 3 involved in vF. neby.

2197 } 2198	2	Each has a nucleus.
		                        Pos. Dist.
		Sept. 29, 1850. βα 178°
		                  βα 178 1' 35"
		                  βα 176 1 34
		                  βα 175 1 34
		The last two observations probably most correct.
		             Position of axis of β 225°
		             Position of axis of α 160

2199	3	Sept. 16, 1854. ✳ in np. edge. ✳ seen in centre of nucleus?
2200	2	F; bM.
2201	2	Aug. 24, 1851. ✳ p. the nucleus; E. np, sf.
2205	5	Since 1850. Nothing further. Nov. 26, 1850. Pos. Dist. βα 23° 1' 46"
2206	2	Looks like a ✳ seen in haze.
2208	1	vF; several ✳s involved.
2209	1	Mottled; ✳ in np. edge.
2210	1	S; lE. p. and f.
2214	3	Nucleus.
2215 } 2216	2	Oct. 9, 1850. Pos. 223°. Dist. 2' 52".
2218 } 2219	3	Nov. 2, 1850. 4 neb. in the field.
2220	1	Nucleus.
2221	1	R; pL.
2222 } 2223	1	Both R, and have nuclei.
2224	3	Sept. 16, 1852. Involves a vS. ✳ to nf. Another neb. 6' p. and 1' n. of it.
2226	2	Oct. 8, 1855. Outline irreg; pB.
2227	1	R.
2228	5	Like 242. Oct. 11, 1850. Much E. from np. to sf; gbM. to nucleus. Sept. 18, 1852. S. ✳ p. nucleus, and on edge of neby.
2230 } 2231	1	Aug. 30, 1851. Another neb. f. 2231 about 12', which is E. p. and f.
2232	7	Oct. 3, 1856. sf. edge is the brighter, and the more sharply defined.

Number in Herschel's Catalogue.	Number of times observed.	Description.
2236		3 or 4 conspicuous *s in it; not in a line between Herschel's two *s, the p. one of which is d.
2237	2	S; vF; R.
2241	16	Since the publication in the 'Transactions' for 1850. The outlying portions in the published sketch are parts of spiral branches. Figure 40, Plate XXX. represents it as seen on a very fine night (Sept. 16, 1852), with a freshly polished speculum which defined very sharply. Oct. 2, 1856. All the details in Mr. STONEY's drawing very well seen. Oct. 16, 1857. The spiral arms and the * in centre distinctly seen.
2242		Oct. 23, 1857. R; pB; nucleus; another 6' s; S; vF.
2245	12	Sketched 4 times. Nov. 5, 1850. I saw two knots and a dark space between them. I think the neb. is connected above the dark space. Nov. 27, 1850. 2 knots seen nearly n. and s, and a dark space between. Aug. 24, 1851. 2 knots and a dark space between, connected above by neby, as in sketch. Sept. 26, 1854. Certainly a spiral; some *s at moments visible. Oct. 17, 1854. Spirality distinctly seen. I thought the coil doubled in upon itself more closely than shown in Mr. STONEY's drawing, and that the central knot had a stellar nucleus. The whole neb. looked sparkling, though I could not see its separate *s. Nov. 22, 1854. Central nucleus stellar. The outer edge of the coil, just where it joins the external nucleus, seems brighter than the rest. Oct. 15, 1855. Seen to be spiral, as before. See fig. 41, Plate XXX.
2248	2	Oct. 8, 1855. eF; mottled and irreg. outline.
2250		Looked for 4 times; not found.
2257	1	Nov. 4, 1850. Nucleus; a F. neb. f. about 2'.
2258	1	Oct. 17, 1854. R; pB; mbM.
2260	1	Nov. 13, 1854. E; bM; a F. * p.
2261	4	Oct. 24, 1857. Edge ragged; F. nucleus.
2262	3	Oct. 24, 1857. pB; R; mbM.
2264	6	Oct. 7, 1855. A F. suspicion of a dark ring round the B. centre.
2267	1	Nov. 17, 1854. lE. n. and s.
2268	5	Nov. 22, 1854. pL; R. A * precedes the nucleus (1-inch single lens); sp. this object there is a vS. E. neb.
2271	4	Aug. 24, 1851. A * with a S. neb. in contact.
2273	1	Oct. 12, 1855. A * p. touches the neb. A little np. is another neb. vvF.
2274 2275	15	3 neb. found Nov. 5, 1850. Pos. Dist. αβ 114° 5' 30" αγ 128 5 34 β 84 1 44
2278 2279 2280 2281	2	All R; gbM.
2282	3	bM.
2284		cl.
2290	6	Oct. 31, 1855. pL; pB; has a F. but pretty sharp nucleus; edges ragged.
2291	2	pB; pmE.
2297	13	Oct. 12, 1855. pL; B; E; gmbM. A decided dark lane runs through it in the direction of its major axis. The neb. is rather narrower in the middle of its length, and spreads out laterally towards its extremities, fading away very gradually. Nov. 3, 1855. Seen as before; dark streak through centre quite plain. Sept. 9, 1856. Seen very well; dark lane through centre quite plain, especially with highest single lens. Oct. 3, 1856. I think I see right-hand side of centre to be composed of *s: It is brighter than the opposite side. See fig. 42, Plate XXX.
2299	2	Oct. 17, 1854. R; pB; bM. to a nucleus.
2300	11	Sept. 10, 1849. 2 S. *s near M. Oct. 7, 1855. No nucleus; 2 *s seen steadily. The centre of neb. looks darker than the rest. Oct. 8, 1855. There certainly exists a dark bay in the centre of the neb. between the two *s.
2301	1	Aug. 24, 1851. A S. lenticular neb.

Measurements of H. 1622 by Mr. BINDON STONEY, October 4, 1851.

H. 1622, April and May, 1851. Object.	Mean of the observation of position.	No. of observation.	Greatest difference between observation and mean.	Mean of the observation of distance.	No. of observation.	Greatest difference between observation and mean.
9 N	49·37	2	1 0	88·20	2	0·3
9 n	22·14	2	0 23	338·7	2	1·2
9 12	323·7	1	......	81·3	1	
9 15	336·37	1	......	187·5	1	
9 5	115·52	2	1 15	107·7	2	0·0
9 4	97·7	1	......	294·9	1	
9 2	45·37	1	......	377·7	1	
9 13	296·7	1	......	177·3	1	
9 14	326·37	1	......	234·9	1	
9 8	217·37	1	......	116·1	1	
9 7	196·37	1	......	93·3	1	
9 6	175·7	1	......	186·9	1	
9 10	201·37	1	......	48·9	1	
9 11	333·37	1	......	57·3	1	
9 16	249·37	1	......	201·9	1	
5 α	......	...	......	254·1	1	
5 β	......	...	......	94·5	1	
15 γ	......	...	......	300·5	1	

Object.	H. 1622. Position as measured by			H. 1622. Distance as measured by		
	O. Struve, 1851.	J. Stoney, 1850.	B. Stoney, 1851*.	O. Struve, 1851.	J. Stoney, 1850.	B. Stoney, 1851*.
N 1	51 47	52 4	......	115·1	126·6	......
N 2	54 48	54 0	44 27	518·0	300·0	289·27
N 3	......	104 20	......	......	165·6	......
N 4	108 54	111 57	112 42	243·63	243·6	243·95
N 5	161 47	165 35	164 11	104·4	103·2	108·67
N 6	190 24	191 42	191 54	250·0	234·0	249·12
N 7	210 51	211 2	212 35	174·2	156·6	174·47
N 8	221 25	220 49	222 43	202·6	176·79	203·57
N 9	229 26	231 32	229 37	88·57	83·4	88·2
N 10	......	223 30	219 37	......	106·94	133·77
N 11	277 27	279 21	266 8	121·9	109·8	92·88
N 12	......	274 23	273 40	......	117·67	116·15
N 13	......	281 37	275 8	......	239·0	227·67
N 14	......	......	305 11	......	......	240·39
N 15	309 2	310 34	308 40	189·9	175·8	183·15
N 16	......	......	243 30	......	......	286·62
N n	14 51	16 54	13 23	265·65	262·2	262·97

* In the reduction of these from the former Table, STRUVE's determination of N 9 has been used.

N.B. STRUVE's measurement of N 11 ought perhaps to be attributed to N 12.

List of Stars in the Dumb-bell Nebula, H. 2060, measured in autumns of 1850 and 1851. Origin taken at α (a of M. STRUVE); brightest star in sp. quarter. By Mr. BINDON STONEY.

Name in observing-book.	M. Struve's name.	X.	Y.	No. of observation.	X in M. Struve's list.	Y in M. Struve's list.
β	d	$+154\cdot4$	$+ 17\cdot5$	2	$+152\cdot5$	$+ 12\cdot8$
ζ	c	$+325\cdot3$	$+ 63\cdot1$	2	$+321\cdot1$	$+ 62\cdot8$
γ	e	$+172\cdot8$	$+ 98\cdot6$	3	$+172\cdot5$	$+ 98\cdot8$
δ	f	$+172\cdot3$	$+144\cdot5$	2	$+174\cdot6$	$+142\cdot2$
ε	g	$+173\cdot9$	$+193\cdot2$	2	$+174\cdot3$	$+191\cdot5$
θ	h	$+246\cdot2$	$+100\cdot5$	2	$+251\cdot1$	$+101\cdot7$
$\varkappa$	k	$+187\cdot4$	$-121\cdot3$	2	$+186\cdot8$	$-121\cdot0$
μ	w	$+323\cdot5$	$-115\cdot0$	2	$+317\cdot7$	$-102\cdot0$
o	b'	$+303\cdot6$	$+183\cdot4$	2	$+308\cdot3$	$+186\cdot3$
ρ	b	$+ 23\cdot0$	$+199\cdot7$	2	$+ 21\cdot6$	$+199\cdot7$
σ	o	$+ 46\cdot8$	$+149\cdot5$	3	$+ 49\cdot6$	$+144\cdot0$
π	p	$+ 48\cdot1$	$+ 61\cdot3$	2	$+ 60\cdot6$	$+ 57\cdot9$
τ	a'	$- 10\cdot9$	$- 59\cdot8$	2	$- 12\cdot3$	$- 57\cdot2$
η	i	$+235\cdot8$	$+ 70\cdot8$	1	$+238\cdot2$	$+ 71\cdot5$
ι	...	$+223\cdot0$	$+ 51\cdot4$	1		
A	...	$+ 54\cdot4$	$- 22\cdot8$	2		
λ	...	$+304\cdot6$	$+ 21\cdot3$	1		
ψ	...	$+ 73\cdot6$	$-136\cdot8$	1		
φ	...	$+ 63\cdot6$	$-122\cdot6$	1		
ω	...	$+142\cdot6$	$+ 60\cdot4$	1		
ν	$\prime$	$+398\cdot2$	$- 91\cdot6$	1	$+396\cdot9$	$- 74\cdot0$

A is the extremity of the bright mass of nebulosity in the sp. quarter.

[*Observations of Stars in the Dumb-bell Nebula, Messier* 27, H. 2060. By M. O. STRUVE.

The following list of observations contains all stars which can be seen, by the Pulkova refractor, in the Dumbell Nebula. Sometimes I had a faint suspicion of some stars more visible on the ground of the nebula, but they were not seen distinctly enough to allow an exact micrometrical determination of their places. It will be interesting to see how many stars more Lord ROSSE's great reflector will show in this nebula; from this comparison some approximate judgment might be formed about the respective space-penetrating powers of the two instruments. My observations extend as far as I was able to trace the nebulosity. I think there is only one star in the lists situated quite out of the nebula. This star was observed merely for the sake of control. Probably Lord ROSSE's large reflector will extend considerably further the boundaries of the nebulosity; but to compare the relative powers of the two instruments, it will be necessary to confine the enumeration of the stars to the same boundaries which are approximately indicated by the stars in the following list. As fundamental point for the triangulation, I selected the star a of the 10th magnitude, situated near the S.W. corner of the nebula. It cannot be mistaken, for it is far the brightest object in the whole nebula. All observations have been made with illuminated wires in the dark field, and with a power of 207.

Observations.

Date.	Object.	Magnitude.	Ang. Pos.	No. of measures.	Distance.	No. of measures.
1851. Sept. 1	a b	b = (11.12)	6·4	3	200·4	3
	b b'	b' = (11)	92·7	3	282·3	3 not accurate.
	a e	e = (11)	60·1	3		
Sept.	a e	....	60·5	3	199·4	6
	c e	c = (11.12)	283·7	3	152·9	6
	a d	d = (11.12)	85·2	3		
	e d	....	193·1	3		
	e f	f = (12)	2·8	4		
	b f	....	110·6	4		
	b g	g = (12.13)	93·4	4		
	e g	....	1·1	4		
	e h	h = (11.12)	87·9	5		
	c h	....	299·0	4		
Sept. 3	e i	i = (12.13)	112·6	4		
	h i	....	203·1	4		
	a k	k = (11.12)	123·0	4		
	e k	....	176·4	6		
Sept. 12	a b	....	6·1	3	201·3	6
	a e	....	60·3	3	198·5	6
	e c	....	103·0	3	151·7	6
	a l	l = (10.11)	116·05	6	$\Delta\!R = 358·7$	10
Sept. 17	e m	m = (12)	168·7	4		
	l m	....	268·9	4		
	l n	n = (12.13)	258·2	4		
	m n	....	180·3	6		
	a o	o = (13)	19·0	4		
	b o	....	156·3	4		
	e o	....	290·2	4		
	a l	....	116·00	4	$\Delta\!R = 359·9$	10
Sept. 22	a p	p = (13)	46·3	4		
	b p	....	164·6	6		
	a q	q = (11.12)	257·9	4	99·8	6
	l r	r = (12.13)	36·2	4		
	e r	....	127·6	4		
	s r	s = (8)	258·3	4		
Sept. 26	l t	t = (12)	24·9	4		
	e t	....	142·4	4		
	s t	....	247·0	5		
	l u	u = (11.12)	118·2	6	77·9	9
	u v	v = (12)	343·3	6		
	l v	....	89·5	5		
	l w	w = (13)	346·3	4		
	a w	....	107·8	4		
	s w	x = (13)	256·3	4		
	l x	....	82·9	4		
	u x	....	37·4	6		
	l y	y = (12)	190·5	4		
	u y	....	256·9	4		
	l z	z = (12.13)	234·6	5		
	k z	....	130·3	4		
Oct. 3	a a'	....	192·1	4		
	q a'	....	113·0	4		
	a s	....	$\Delta\delta = -42''·0$	8	$\Delta\!R = 600·8$	16
	a b	....	6·0	4	201·2	8
Oct. 20	a e	....	59·9	3	198·5	7
	e b'	....	57·2	4	161·5	6
	e c	....	104·1	3	154·1	7
Dec. 17	c d	....	253·8	5		
	d k	....	166·1	4		
	e l	'.........	148·72	4		
	l u	....	119·5	4	79·9	6

N

From the preceding observations the following coordinates, X, Y, with regard to a have been deduced. X is the coordinate in the direction of $\mathbb{R}$ or $\Delta \mathbb{R} \cos \delta$; Y is the coordinate in declination or $\Delta \delta$. In this deduction, where the angles of position of any star were taken from three different points previously fixed, I have combined only those two directions which promised the most exact evaluation of the coordinates. Thus, for instance, the star r is related to l, e and s. In the three triangles formed in this way, the angle at r is found respectively $l\,r\,e = 91° 24'$, $l\,r\,s = 137° 54'$, $s\,r\,e = 130° 42'$. Hence it is evident that the nearly right-angular intersection of the two sides $l\,r$ and $r\,e$ in the triangle $l\,r\,e$, must give the best combination for the deduction of the position of r. The relation of l and e to another and to a being previously established, the coordinates of r to a were deduced from the resolution of the triangle $l\,r\,e$. The difference of these coordinates from the corresponding previously fixed coordinates of s with regard to a, gave the angle of position of r from s by calculation $= 258° 37'$. This calculated angle, compared with the directly observed position $= 258° 18'$, shows a difference of only $19'$, and bears, therefore, a very satisfying testimony to the exactness of the deduced position of r. The same proceeding having been followed in all other cases, where control observations had been made, I have got the conviction that the probable error in any of the following coordinates will be considerably inferior to one second of space. It will be comparatively smaller for the brighter stars, but somewhat greater for the very faint objects.

Calculated Coordinates of the Observed Stars with regard to a.

	X.	Y.
q	$- 97 \cdot 6$	$- 20 \cdot 9$
a'	$- 12 \cdot 3$	$- 57 \cdot 2$
b	$+ 21 \cdot 6$	$+ 199 \cdot 7$
o	$+ 49 \cdot 6$	$+ 144 \cdot 0$
p	$+ 60 \cdot 6$	$+ 57 \cdot 9$
d	$+ 152 \cdot 5$	$+ 12 \cdot 8$
e	$+ 172 \cdot 5$	$+ 98 \cdot 8$
g	$+ 174 \cdot 3$	$+ 191 \cdot 5$
f	$+ 174 \cdot 6$	$+ 142 \cdot 2$
k	$+ 186 \cdot 8$	$- 121 \cdot 0$
n	$+ 224 \cdot 9$	$- 184 \cdot 7$
m	$+ 225 \cdot 0$	$- 164 \cdot 3$
i	$+ 238 \cdot 2$	$+ 71 \cdot 5$
h	$+ 251 \cdot 1$	$+ 101 \cdot 7$
z	$+ 279 \cdot 6$	$- 199 \cdot 7$
b'	$+ 308 \cdot 3$	$+ 186 \cdot 3$
w	$+ 317 \cdot 7$	$- 102 \cdot 0$
c	$+ 321 \cdot 1$	$+ 62 \cdot 8$
y	$+ 321 \cdot 8$	$- 219 \cdot 5$
l	$+ 332 \cdot 4$	$- 162 \cdot 2$
t	$+ 347 \cdot 8$	$- 128 \cdot 9$
v	$+ 390 \cdot 7$	$- 161 \cdot 7$
r	$+ 396 \cdot 9$	$- 74 \cdot 0$
u	$+ 402 \cdot 4$	$- 200 \cdot 8$
x	$+ 442 \cdot 0$	$- 148 \cdot 5$
s	$+ 555 \cdot 8$	$- 42 \cdot 0$

Pulkova, Feb. 4, 1852.]

(Signed)   OTTO STRUVE.

[*Observations of Stars in the Spiral Nebula.* H. 1622.

The spiral form of this nebula is very distinctly seen in the Pulkova refractor. Unfortunately in the month of March, the best season for the observation of this object, the sky was constantly cloudy; so that I could only get three nights' observations in the months of April and May, when the twilight did not cease for the whole night. It must be attributed to this unfavourable circumstance that the following list of determinations is not so complete as it probably would have been without the twilight. The observations have been made alternately with powers of 138 and 207.

Observations.

Date.	Object.	Magnitude.	Ang. Pos.	No. of measures.	Distance.	No. of measures.
1851, April 7.	N n	....	14 55	5	267·1	4
	N a	a = (11)	229 24	3	88·0	3
	N b	b = (11.12)	109 12	3	242·6	3
	$a b$	....	93 42	3	298·6	3
April 28.	$a b$	....	94 23	3	300·8	4
	N a	....	228 36	4		
	N b	....	108 54	4		
	$n a$	....	203 42	3		
	$n b$	....	153 30	3		
	$a d$	d = (12.13)	323 51	3		
	N d	....	277 27	3		
	$a e$	e = (13)	112 13	3		
	N e	....	161 56	3		
	N f	f = (12.13)	309 18	3		
	$n f$	....	237 31	3		
	$a f$	....	335 23	3		
	$a g$	g = (12.13)	215 17	3	115·5	4
	$a h$	h = (12.13)	193 29	3		
	$g h$	....	87 5	3		
May 3.	N k	k = (13.14)	51 47	3		
	$n k$	....	173 29	4		
	$b k$	....	317 23	3		
	$b l$	l = (11.12)	27 20	4		
	$n l$	....	83 17	4	355·2	4
	$a e$	....	112 56	4		
	N e	....	161 39	3		
	$a m$	m = (12.13)	172 43	5		
	N m	....	190 44	4		
	$b m$	....	238 50	4		
	N a	....	229 12	4	87·0	3
	N n	....	14 47	4	264·2	3

The results of these measures were deduced in this way, that I first fixed the relations between the four principal objects, namely, the centres of the two nebulæ N and n, and the two brightest stars a, b. In the triangle N $a b$, all distances and directions have been measured It is therefore over-determined, and the definite relations had to be deduced by the calculus of compensation. This calculus gives—

	Ang. Pos.	Distance.
N a	229 26	88·57
N b	108 54	243·63
$a b$	94 6	298·55

To these relations must be added the mean value of the two observed relations N n (April 7, May 3) which gives N n ang. pos. $=14°$ 51', distance $=265''\cdot65$.

These relations between the four principal objects form the base from which the places of the other stars are deduced by resolutions of the triangles formed by the observed directions. The following Table contains the results of my calculations. It gives the places of all observed objects with regard to N, the apparent centre of the greater nebula, and that of the star g to h.

	Ang. Pos.	Distance.
N n	14 51	265·65
N a	229 26	88·57
N b	108 54	243·63
N d	277 27	121·9
N e	161 47	104·4
N f	309 2	189·9
N h	210 51	174·2
N g	221 25	202·6
N k	51 47	115·1
N l	54 48	518·0
N m	190 24	250·0
h g	267 5	44·7

Our observations contain several controls, by which it is proved that the deduced places of the stars, with regard to N, might be judged all exact within 2''. This exactness must be regarded as very satisfactory, if we consider the extreme faintness of the observed objects. I estimate a star to be of the 14th magnitude if it is more suspected than distinctly seen in a dark night. Hence it follows that the greater part of the stars in our list are close to the extremity of measureableness in the Pulkova refractor. Another cause which troubles the agreement of results is the indistinctness of the centre of the greater nebula N. The centre of the small nebula n is much more distinct: all observations of the dimensions of the nebula, or of knots in it, have been omitted by me, as they can be observed with much more accuracy by Lord ROSSE's powerful telescope.

The Earl of ROSSE communicated to me the following relations of sixteen objects in the nebula, as observed through his telescope. I add to that list the differences of our measures (S—R) for the objects which appear to be identical.

Designation by Lord Rosse.	Designation by O. Struve.	Ang. Pos.	S—R.	Distance.	S—R.
N n	N n	16 34	−1 43	4 22·2	+ 3·4
N 1	N h	52 4	−0 17	2 6·6	−11·5
N 2	....	54 0	....	5 0·0	
N 3	....	104 20	....	2 45·6	
N 4	N b	111 57	−3 3	4 3·6	0·0
N 5	N e	165 35	−3 48	1 43·2	+ 1·2
N 6	N m	191 42	−1 18	3 54·0	+16·0
N 7	N h	211 2	−0 11	2 36·6	+17·6
7, 8	$h g$	270 42	−3 37	0 34·8	+ 9·9
N 9	N a	231 32	−2 6	1 23·4	+ 5·2
9, 10	....	197 57	....	0 27·0	
N 11	N d	279 21	−1 54	1 49·8	+12·1
11, 12	....	225 27	....	0 12·6	
N 13	....	281 37	....	3 51·0	
14, 15	....	297 15			
N 15	N f	310 34	−1 32	2 55·8	+14·1

From this comparison it is evident that all angles measured by Lord ROSSE are too great. The mean value of the correction, 1° 57′, corresponds to a linear distance of −4″·7. The distances appear to be generally too small. The mean value of the differences is +6″·8. Perhaps Lord ROSSE's star 2 is identical with my l; but in that case, Lord ROSSE's distance N 2 must be an error of writing. At the distance of 5′ from N I could not see the least trace of a star in the indicated direction. In my copy of Lord ROSSE's diagram the star 2 is placed at a distance of about 8′, corresponding with my observations; 10 appears to me only as a knot of the nebula, and has therefore not been measured by me. About the stars 3, 12, 13, and 14 there is no notice given in my journal. Perhaps they might be seen and measured with our refractor. The next spring I intend to repeat and to complete the series of observations, and to decide on the visibility of the not yet noticed stars.

(Signed)   OTTO STRUVE.

(Dated) *Pulkova, June* 2, 1851.]

In forming some estimate of the degree of reliance to be placed on the micrometrical measurements in this paper, we have taken advantage of the information so obligingly communicated by M. O. STRUVE.

As to the measures of distance, they accord with STRUVE's as closely perhaps as could be expected. We measure with bars instead of lines, and without illumination, that we may the better see the faint details of the outlying portions of the nebulæ; besides, we do not employ clockwork to move the telescope. Our measures of *stars* cannot therefore in accuracy compete with STRUVE's, but they are quite sufficient for giving precision to the drawings. As to the angles of position, the same remarks apply, with this addition, that we refer our measurements to the horizon, and reduce them to the equator. Our zero is therefore obtained from the spirit-level, which saves time, to us a great object. We proceed in this way: the level is made horizontal and read off: each measure + or − this quantity will be the distance from the horizon, provided the telescope is on the meridian. When the measures are not taken on the meridian, but a little before or after it, there will be an error in all positions except at the equator. There will be an error owing to two causes, one the error of the zero, the level having been read off the meridian; the other the error in the reading of the position-circle, owing to the action of the universal joint which carries the telescope. The transverse axis of this joint restrains the movement of the telescope round the line of collimation as it approaches or recedes from the meridian; and consequently the plane of the position-circle, except at the equator, does not pass through the pole. The sum of the errors in each case could easily have been computed and allowed for in the reductions, had the distance from the meridian been taken simultaneously with the measurements, but this would have taken much time.

The measurements in the Dumb-bell accord pretty closely with STRUVE's, and may, I think, be taken as a fair average of the work.

As to H. 1622, the comparison of STRUVE's measures with those of the two STONEYS will give more than a probable amount of error at 42° N.P.D, because the stars are numerous, and some measures therefore were taken at a considerable distance from the meridian.

I have not seen the Dumb-bell since STRUVE's letter, having been from home when it was within reach; and no attempt has been made to ascertain the largest number of stars visible in it. No stars have been inserted in the sketch which have not been measured: very many more were distinctly seen. The number of stars visible in this nebula depends even more upon magnifying power and distinctness than aperture; high powers obliterate the faint nebulous details.

The only additional information as to the limits of the nebula which has been obtained since Mr. B. STONEY's drawing was made is contained in the following entry:—August 29th, 1854, observed by Dr. ROBINSON and Mr. J. STONEY. Mr. STONEY says, "Both Dr. ROBINSON and I agreed that the band of faint nebulosity extended further down than in my brother's drawing. My brother and I had formed the same opinion on a previous occasion."

In the observations a silver speculum is sometimes mentioned: we have employed silver occasionally for the second reflexion in this way. First, a *thin* deposit on glass by LIEBIG's process. This, even when fresh, reflects but little more light than speculum-metal. Second, a *thick* deposit on glass, by the grape-sugar or tartaric-acid process, transferred to brass by a thin film of shell-lac: this reflects much more light; but the manipulation is rather difficult, and the surface is not very durable. Third, a surface of standard silver, polished by mechanical means. Fourth, parallel glass, silvered by the grape-sugar process; this of course is durable, but very inferior to the uncovered silver in light and in definition. These substitutes for speculum-metal have only been occasionally used, and for special purposes.

List of Nebulæ not found.

Number in Herschel's Catalogue.	Number of times looked for.	Observations.	Number in Herschel's Catalogue.	Number of times looked for.	Observations.
57	1		672	1	
162	8		706	6	
184	1		745	1	
206	1		828	1	
281	1		1307	1	
284	2	One night passing clouds.	1485	2	Once sky hazy.
314	1	Clouds passing.	1535	1	Clouds passing.
333	1		1832	1	
343	7		1948	1	
356	4	{ Twice, a slight milkiness suspected.	1974	1	
			2062	1	
401	9		2073	2	
468	5	No nebulosity seen.	2113	2	
546 } 577 }	1		2133	4	
			2137	1	
578	1		2148	3	Once clouds passing.
590	1		2250	4	
669	1		2302	2	

This is not to be considered as a list of missing nebulæ, but merely of objects which were not found in the ordinary course of observing, and to which therefore it is desirable that attention should be directed. They have not been looked for since this list was made out.

The engravings of the Nebulæ are extremely faithful: there is, however, a slight inaccuracy which it is necessary to notice, and for which we are to blame, not the engraver. Many of the principal stars are too large. The error arose in this way. The stars were inserted in common, not Indian ink, and, the drawings during their transmission by post becoming slightly damp, the ink made its way into the paper, the dots in some cases becoming small blots. In a few instances it was necessary to set this right to prevent misconception, and some alteration was in consequence made in the Plates; but as to the remainder, we thought it sufficient to state the fact generally, that many of the principal stars were somewhat too large. This remark applies especially to figures 1 and 3, Plate XXV., and to figures 24, 28, 29, 30, 34, Plate XXVIII.

III. *An Account of the Observations on the Great Nebula in Orion, made at Birr Castle, with the 3-feet and 6-feet Telescopes, between* 1848 *and* 1867. *With a drawing of the Nebula. By Lord* OXMANTOWN. *Communicated by the Earl of* ROSSE, *K.P., F.R.S.*

Received June 17,—Read June 20, 1867.

MANY drawings of the Great Nebula in Orion have already been published by different astronomers; still as the present drawing was made with the advantage of instrumental power far exceeding that at the disposal of previous observers, and as great care has been taken to make it accurate, in fact every available hour during the winter months of seven seasons having been employed upon it, perhaps it may be of some interest to the Royal Society.

Several drawings of this wonderful object were published previous to the year 1825, but they were made with instruments of little power; however, in 1825, Sir J. HERSCHEL published a drawing made with his celebrated 18-inch reflector (Memoirs of the Astronomical Society, 1826). Sir J. HERSCHEL's second drawing with the same instrument, but under more favourable circumstances, together with a description and a catalogue of stars, was published in 1847* (Cape of Good Hope Observations). That was succeeded in 1848 by Mr. BOND's drawing, also with a description and catalogue of stars (Memoirs of the American Academy of Arts and Sciences, 1848).

In 1854 Mr. LASSELL published his observations on the Nebula in Orion made at Malta in 1853 (Memoirs of the Royal Astronomical Society, 1854); and in 1862 MM. O. STRUVE and LIAPOUNOV's memoir, with a catalogue of stars, was published at St. Petersburg.

The observations upon this nebula, recorded in the Journal of the Observatory at Parsonstown, date from 1849. From that time till February 1858 there are entries of 54 observations.

In the year 1852 Mr. BINDON STONEY made a drawing of the Huygenian region, which is a very interesting record. Mr. BINDON STONEY was a highly educated civil engineer, well accustomed to use his pencil; he accepted the office of assistant till he was enabled to obtain an engineering appointment. His drawing was made with great care, and he was engaged upon it the whole season. It was compared by several persons with the nebula, and was considered exact. When we compare this drawing with the nebula as it is at (Plate I.) present, there are strong indications of change

---

* This drawing appears to have been executed principally during the years 1835, 1836 & 1837.

THE HUYGENIAN REGION OF THE GREAT NEBULA IN ORION

Between February 1860 and February 1864 there are 74 entries of observations. In February 1860, Mr. HUNTER, who was then the assistant, being an accomplished artist, commenced a new drawing, and was engaged upon it till February 1864, when he was obliged to retire from ill health.

As a groundwork for his drawing, Mr. HUNTER laid down all the stars given in ' Observations de la grande Nebuleuse d'Orion faites à Cazan et à Pulkova, par O. STRUVE, St. Pétersbourg, 1862,' in the positions given at page 118 of that treatise, the nebulosity was gradually filled in by eye as correctly as possible with reference to the stars given in that memoir, and twenty-eight* additional stars from the 9th to the 15th magnitude were inserted by eye-estimation.

During the season 1864–65 the nebula was often examined with the view of verifying the drawing made by Mr. HUNTER, and in 1865–66 some additions were made to it in the neighbourhood of the stars 1_{ix}, 2, 3, 4, 1_{vii}, 2_{i}, 9_{i}, 27_{i}, 45_{i} (Plate II.); also the positions of a few stars not given in STRUVE's list were determined by rough micrometrical measurement, and others were laid down by eye-estimation. During the season 1866–67 these measures were completed, the additions of the previous season verified, and the drawing, extended to the neighbourhood of the stars 1_x, 0_{vii}, 0_{viii}, 46_i, 99_i, 143, 149, and 147.

The facts which seem to be of the most interest have been collected under the following heads :—

1st. List of stars which do not occur in OTTO STRUVE's memoir.

2nd. Distance to which the nebulosity extends in various directions.

3rd. Peculiarities of form in some parts.

4th. Evidence of variability in some of the stars.

5th. Evidence of change in the form and brightness of various part of the nebula.

6th. Resolvability of parts of the nebula.

7th. Spectrum of the nebula.

List of New Stars.

The following is a list of the stars which have been inserted in the drawing, but which are not contained in the list at page 118 of M. O. STRUVE's memoir.

In order to avoid altering Sir J. HERSCHEL's numbers and at the same time to keep the whole list of stars in regular order in right ascension, each of the stars in this list has affixed to it the same number as that of the star in HERSCHEL's list which next precedes it in R.A., one, two, three, &c. dots or Roman figures being placed under the number to distinguish each from HERSCHEL's star and from all other new stars. Some of these stars have been laid down by rough measures of their distance and position with respect to the nearest stars in STRUVE's list, but the greater number have been laid

---

* Three of these are not inserted either in the engraving or in the list of stars, as I was unable to find them in Mr. HUNTER's drawing. The positions given in Mr. HUNTER's list are $-1'\,5''$, $+0'\cdot52$; $+2'\,30''$, $+1'\,35''$; $7'\,13''$, $-3'\,20''$

down by eye-measurement. The measures were taken with a wire micrometer without illumination, the parallel wires being bars $\frac{1}{10}$ inch in breadth, formed each of four wires laid side by side; the position wire was a silk thread about $\frac{1}{40}$ inch thick. The instrument generally used for these measures was an equatoreal of 18 inches aperture and 10 feet focal length, driven by a water-clock movement, as described in the Monthly Notices of the Astronomical Society for 1866. Two measures only were taken, when they were found to agree pretty well; the object being, not to place the stars with the utmost accuracy, but to lay them down so near their true position that the drawing of the nebula might not be distorted to any appreciable extent.

Those stars which are marked (M) have been determined by micrometrical measurement, the remaining ones by eye-estimation. The stars marked (H) have been inserted by Mr. HUNTER.

This list does not contain all the stars which can be seen within the limits of the drawing. The positions given in the following list are taken from the accompanying Skeleton Map by the method which HERSCHEL describes (Cape Observations, Sect. 58).

	ΔR from 69.		$-\Delta$ Declination.			Approximate magnitude.	
0_I	-36	40	-15	55	M	10	Distance only from 1_{VII} measured.
0_{II}	-36	25	$+ 6$	40	...	10	
0_{III}	-36	0	$+13$	30	M	10	
0_{IV}	-35	20	-17	20	...	13	
0_V	-34	55	-17	40	...	13	
0_{VI}	-34	55	$+ 7$	35	...	11	
0_{VII}	-33	45	$- 4$	15	...	12	
0_{VIII}	-33	20	-11	30	...	9	
1_I	-30	30	$+ 3$	35	...	13	
1_{II}	-29	5	$- 4$	50	...	11	
1_{III}	-28	40	$+ 5$	25	...	11	
1_{IIII}	-27	55	$+ 1$	45	...	10	
1_V	-26	55	$+14$	0	...	13	
1_{VI}	-26	45	$- 1$	20	...	12	
1_{VII}	-26	55	-22	25	M	10	
1_{VIII}	-26	30	-39	20	M	8	
1_{IX}	-26	10	-10	0	...	12	
1_X	-26	10	$+17$	10	...	12	
1_{XI}	-25	5	$- 3$	30	M	10	
1_{XII}	-23	0	$+ 8$	0	...	13	
1_{XIII}	-23	0	$- 7$	30	...	13	
1_{XIV}	-23	0	$+13$	20	M	8	
2_I	-22	5	-22	55	M	9	
2_{II}	-21	10	-13	55	...	14	
2_{III}	-21	0	-19	45	...	14	
2_{IV}	-20	45	-18	50	...	14	
4_I	-20	0	$+ 5$	45	H	9	
4_{II}	-18	50	-12	20	...	15	
4_{III}	-18	20	-13	5	...	14	
4_{IIII}	-18	10	-13	40	...	13	
4_V	-17	30	$+12$	30	H	9	
4_{VI}	-17	5	$+ 8$	55	H	12	
9_I	-14	30	-29	10	...	12	
13_I	-12	20	$+ 8$	15	H	12	
16_I	-12	5	$- 1$	55	H	13	
16_{II}	-12	0	$+10$	35	H	13	

List of New Stars (continued).

	Δ Æ from 69.		−Δ Declination.			Approximate magnitude.	
18$_\text{I}$	−11	30	−18	30	...	14	
25$_\text{I}$	− 9	20	−36	5	...	9	
27$_\text{I}$	− 9	5	−27	40	M	12	
28$_\text{I}$	− 8	50	−18	50	...	14	
28$_\text{II}$	− 8	45	−18	0	...	14	
29$_\text{I}$	− 8	5	−20	0	...	14	
29$_\text{II}$	− 8	10	−36	30	M	7	
29$_\text{III}$	− 8	0	−37	0	...	7	
30$_\text{I}$	− 8	45	−23	10	M	11	
32$_\text{I}$	− 8	15	+12	15	...	12	
33$_\text{I}$	− 7	40	+11	45	...	12	
34$_\text{I}$	− 7	10	−29	20	M	12	
34$_\text{II}$	− 6	40	−21	50	M	11	
36$_\text{I}$	− 6	15	−20	10	...	14	
36$_\text{II}$	− 6	5	−19	15	...	13	
37$_\text{I}$	− 5	50	−35	40	...	9	
37$_\text{II}$	− 5	40	−20	40	M	11	
37$_\text{III}$	− 5	50	−39	10	...	9	
45$_\text{I}$	− 3	45	−20	5	...	14	
46$_\text{I}$	− 3	45	−37	20	M	5	
46$_\text{II}$	− 3	20	−36	45	M	5	
47$_\text{I}$	− 3	10	−35	50	...	9	
51$_\text{I}$	− 2	10	−28	20	M	8	
51$_\text{II}$	− 1	55	−32	0	M	7	
52$_\text{I}$	− 1	25	−38	25	...	9	
56$_\text{I}$	− 1	0	− 2	40	H	13$\frac{1}{2}$	
58$_\text{I}$	− 0	45	+ 1	30	H	14	
58$_\text{II}$	− 0	30	+ 3	50	H	14$\frac{1}{2}$	
59$_\text{I}$	− 0	40	−26	10	M	7	
61$_\text{I}$	− 0	20	−28	10	...	11	
69$_\text{I}$	0	0	−38	25	M	7	
75$_\text{I}$	+ 0	25	+ 1	20	H	15	
88$_\text{I}$	+ 1	0	−27	30	...	11	
88$_\text{II}$	+ 1	15	− 3	35	H	14	
91$_\text{I}$	+ 1	38	− 0	42	H	13$\frac{1}{2}$	
91$_\text{II}$	+ 1	42	− 0	20	H	13$\frac{1}{2}$	
96$_\text{I}$	+ 2	0	−25	15	M	7	
96$_\text{II}$	+ 2	10	−27	40	M	8	
98$_\text{I}$	+ 2	15	−26	0	...	10	
99$_\text{I}$	+ 2	20	−30	55	...	3$\frac{1}{2}$	Taken from B. A. Catalogue = ι Orionis.
100$_\text{I}$	+ 2	25	− 7	0	H	13$\frac{1}{2}$	
112$_\text{I}$	+ 4	0	− 5	50	H	13$\frac{1}{2}$	
112$_\text{II}$	+ 4	5	− 7	0	H	13$\frac{1}{2}$	
113$_\text{I}$	+ 4	40	− 5	45	H	13$\frac{1}{2}$	
113$_\text{II}$	+ 4	45	+ 7	5	H	14$\frac{1}{2}$	
113$_\text{III}$	+ 4	45	+ 7	30	H	15	
114$_\text{I}$	+ 4	25	−37	0	...	11	
115$_\text{I}$	+ 5	20	−40	25	...	8	
125$_\text{I}$	= V		......		...	......	Struve.
127$_\text{I}$	+ 7	25	+ 2	40	H		
129$_\text{I}$	+ 7	40	+ 0	30	H	15	
129$_\text{II}$	+ 7	50	−34	10	...	8	
135$_\text{I}$	+ 9	40	− 3	25	H		
135$_\text{II}$	+10	20	− 2	10	H	14	
138$_\text{I}$	+11	10	+ 0	20	H	14$\frac{1}{2}$	
138$_\text{II}$	+12	10	−35	20	...	8	
142$_\text{I}$	+14	20	− 0	45	H	14$\frac{1}{2}$	
150$_\text{I}$	+23	10	− 5	50	...	9	

Remarks on the foregoing List.

Of the existence of all the stars in the foregoing list which are not marked with an H I am fully satisfied, and of the remainder I have seen 4_v, 4_{vi}, 13, 16_i, 16_{ii}, 56_r.

The nebulosity I examined very carefully, and except in a few places I was able to verify most of the details. The three rays of nebulosity between the stars 13_i, 16_{ii}, 28, 39 (Plates II. & III.) I was never able to see well, and on only one occasion did I succeed in seeing two of them*. The patches of nebulosity in the region included between the stars 25, 56, 79, and 99 I never succeeded in seeing, nor did the rays in the region between 133, 142_i, and 149 come out as clearly as represented in the drawing; the greater part of the nebulosity on the south and preceding sides of the drawing is very faint and ill defined, but the curved outline above 20 is tolerably sharp, and the neighbourhood of 20 and 30_i is darker than anywhere in the surrounding parts, even as far as 99_i, 37_{iii}, and 1_{vii}.

Extreme limits of Nebulosity.

The distance to which the nebulosity can be traced is very great, but it fades away so very gradually that it is quite impossible to assign any definite limit to it.

M. LIAPOUNOV, with an achromatic telescope of $9\frac{1}{2}$ inches aperture, was able to trace it up to a line a little beyond 15 through 34, 74, 135, 126, a little outside 135_i near 82, 46, 6, 4_{vi}, and 5, and the spiral round 108†.

Mr. BOND does not mention the extreme limits to which he was able to trace the nebulosity except in two directions. He says, " The connexion of the main body of the nebula with that portion which surrounds C Orionis is traced by the north preceding route. It is quite decided," &c. He also says that the main body of the nebula is not connected with C on the following side; and again, " I was unable to satisfy myself how far it might be possible to trace it (the light) southward, but certainly beyond *ι*. Soon after this star it, however, becomes very faint"‡.

His drawing extends to the limits mentioned by LIAPOUNOV, and therefore he was probably able to trace it rather further on all sides.

Mr. LASSELL, with his reflector of two-feet aperture, did not extend the nebula quite as far as Mr. BOND in his drawing, but probably he was able to trace it a good deal further.

Sir J. HERSCHEL's drawing in the Cape Observations extends to the stars 6, 124, 145, 141, 96_i, a point halfway between 5 and 1_{xi}, and to 34 and 58, and. he says (page 29), " The area of the figure (half a square degree in extent) comprises all the nebulous convolutions and appendages which I have been able to trace, with the exception of the terminal effusion of the greater proboscis beyond the star A ($=135$) southwards, which may be traced as far as the double star *ι* Orionis, which it involves and renders nebu-

---

* Mr. HUNTER during the last season sometimes employed the 6-feet telescope as a Herschelian; I never did so; silvered glass was, however, occasionally substituted for the flat mirror of speculum-metal, which diminished the loss of light by the second reflexion, but it soon became dewed and the silver surface broke up.

† STRUVE's Memoir, page 2.

‡ BOND's description of the nebula about θ Orionis, page 93.

lous. It is, however, of little intensity, and offers nothing remarkable enough in respect of form to make it worth while to enlarge the dimensions of the engraving sufficiently to include the whole."

He also says (page 29), " Northwards between this nebula and C Orionis;   no nebulous connexion has been traced."

With regard to the extent to which the nebulosity can be traced with the 6-feet telescope I do not find any record by Mr. BINDON STONEY. He appears to have confined his attention entirely to the central portions, except in the following instance, where he remarks—

" Nov. 25th, 1851. There is a long dark channel following the Huygenian region by about the diameter of the latter, in which no nebulosity exists, and twice on good nights with a freshly polished speculum it gave me the idea of immense depth, like a gulf without any bottom; ordinarily the light from the surrounding parts spreads a faint light over it. Comparatively few stars follow this part of the nebula for about 80'."

With reference to the extent of the nebulosity, Mr. HUNTER says, " It has been repeatedly traced up to the star ι Orionis on the south, and on several occasions to C Orionis on the north, while in the preceding direction the sky assumes a peculiar milkiness, at least one degree before the nebula comes into the field. In the following direction it does not seem to extend much (about 10') beyond the limits in the sketch."

" Between the stars 135 and ι Orionis the nebulosity narrows to a band of about 5' in breadth, and then again expands as it approaches ι; there has been no attempt to trace it further in this direction."

" Again, the nebulosity curves round from the star 6 in a north preceding direction until it joins a narrow band of faint nebulosity, passing in the preceding and following direction through the little group of stars of which C Orionis is the brightest."

On many occasions I have examined the neighbourhood of this nebula with the view of determining, as far as possible, the extent of the nebulosity. On the following side I was able to trace it 35' following the trapezium; this nebulosity was, however, excessively faint, and of almost uniform intensity; in fact the only proof of its existence was the prolongation of the dark lane extending through it from the star $142_{\mathrm{\scriptscriptstyle I}}$, which made the surrounding region look slightly luminous by contrast. On the north side the nebulosity appears to be nearly cut off short by the same dark lane which extends by the stars 114, 56, 25, 1_{xiv}. At the last star this ceases, and a faint broad streak of nebulosity appears to curve round in the direction of C Orionis, nearly in the position described by Mr. HUNTER. On the north side of this lane the nebulosity seems to begin again, and gradually increases in intensity to the streak at C Orionis, the brightest part being perhaps rather fainter than the nebulosity round ι.

On the preceding side the nebulosity appears to extend to a great distance, and seems to be of the same character as that about the stars 0_{iv}, 0_{viii}, 1_{ii}; the streaks, however, gradually diminish in intensity, and though, when followed by the eye in the preceding direction, they may be traced for some distance (5', or in some cases perhaps 10') beyond the drawing, yet when once lost sight of they cannot be recovered again without returning to the brighter parts and tracing them in the preceding direction again. The great difficulty in tracing them, however, appears to be due, not so much to their extreme faintness, as to the uniform intensity of the nebulosity in this region; for when the nebula is examined with the equatoreal of 18 inches, a very low power capable of taking in a field of 1° being used, the whole sky in this region has a general luminous appearance when contrasted with the following and north following region, which itself is not entirely free from nebulosity. In drawing the greater part of the outlying regions of the nebula an eyepiece of a magnifying power of about 230 only, and having a field of 26′, was generally used; higher powers do not seem to show as much as the lower power, although with the higher powers the whole of the light from the large mirror is received into the eye, which is not the case with the lower power. With higher powers the field does not appear to be sufficiently large to allow of each streak of nebulosity being compared with the surrounding parts where the nebulosity is not quite so bright. The higher powers, however, bring out minute stars with ease which are hardly visible with the lower power. On the south preceding side the nebulosity appears to be much of the same character as on the preceding side; long streaks have been traced from 1_{VII} and from a point about 3′ or 4′ south of it, which have been represented in the drawing just at their commencement. Another streak extends from the south side of 9_I through 46_I, 46_{II}, which seems to extend considerably further than the drawing,—also another from ι in a south following direction.

It is probable that the drawing might be extended considerably further in various directions, but the nebulosity is of such extreme faintness that the work would advance very slowly, the eye requiring so long, after each exposure to even very feeble lamp-light, to recover its full power.

Form.

Very little need be said on this subject as the drawing will speak for itself; it may, however, be well to call attention to the apparent connexion between some of the stars and the nebulosity near them.

In some places the stars appear to have either repelled or absorbed the nebulosity, for instance at the trapezium, at 32 and 35, and 80; and in other places the nebulosity is denser, as if the stars had attracted it, for instance at 2_{II}, 4, 34, and 108. Around the star 108 the nebulosity seems to have a spiral character, and the same appearance, though much less decided, may be seen round 4. Round the stars 46_I, 46_{II}, and 99_I the nebulosity seems to have been concentrated, but close to them there appears to be an absence of nebulosity; and in the case of 99_I the dark hole is situated excentrically with respect to the principal star, its nearer companion being close to the opposite side of the hole*; but in the case of the double star 46_I, 46_{II} the hole is nearly symmetrically situated, but the nebulosity is brightest at the north preceding side. We can

---

* A drawing of the nebulosity around ι was published in the Philosophical Transactions for 1850, in which the hole is well shown.

hardly, therefore, account for these numerous coincidences, except by supposing some at least of the stars to be situated nearly at the same distance from us as the nebula—in fact immersed in the nebulous matter.

Evidence of Change.

1. *Variability of the Stars.*—The only remark concerning the variability of a star which I find in Mr. Bond's paper (page 94), refers to star 78, which he thinks is a variable star of short period.

In Mr. Lassell's paper I do not find anything on this subject.

A star marked x in Sir J. Herschel's diagram, published in 1825, was not found by him in 1837; perhaps this may be the same as 129_1, but as there is no list of positions of the stars in the paper of 1825, the identity is uncertain. Possibly this star may be variable.

Although M. Liapounov's observations were made about fourteen years after Sir J. Herschel's, Struve found that the magnitudes of the stars as given by Herschel agreed very fairly with his (Liapounov's) determinations, but one star (IV.), whose position was estimated by M. Liapounov, and at about the same time was measured by Mr. Lassell, does not occur in Sir John Herschel's list, although it is so situated that it would not be likely to be overlooked. M. Struve examined with considerable care Bond's Catalogue of Stars in this nebula with the view of identifying them, as far as possible, with those in Sir J. Herschel's list. B 26 and B 27 he considers identical, and they are near the position of 75, but Mr. Bond appears satisfied of the existence of two stars here of the 17th and 18th magnitudes respectively; their positions are only about $2\frac{1}{2}''$ apart, and therefore their combined light might easily be taken for that of one of $12\frac{1}{2}$ magnitude, which is that assigned to 75 by Struve; Herschel, however, assigns to 75 the 18th magnitude, so that the star is possibly variable.

Mr. Bond gives two stars about $11''$ apart near the position of 57, but no one else seems to have seen more than one. I have examined this part of the nebula repeatedly, but have never seen more than one star in this position. He also gives two stars south following 88, which may be identical with the two stars 91 and 91_1. Mr. Hunter gives the two stars in his drawing, but he is not quite sure whether 91_1 is not the same as 91, as he was not able to find the latter on the night on which he saw 91_1.

Bond's stars 42, 85, and 88 were not found by either Struve or Hunter; there are also several others which Struve could not identify and which are not given by Hunter, but Struve is of opinion that some errors exist in Bond's positions.

Struve next examines Lamont's catalogue of 34 stars, all of which he identifies more or less satisfactorily with stars of Herschel's list, with the exception of four, viz. 4', 5', 7', and 28*; 4' and 7' I do not find in Hunter's list; 5' is in very nearly the position of 75_{II} in Mr. Stoney's drawing† (I do not find any mention of this star in Mr. Hunter's observations, but I am almost certain that I have seen one in this position and on the edge of the nebulosity); 28 is not more than $15''$ distant from Struve's star II.

---

* See Struve's Memoir, p. 95; these numbers are Lamont's.   † See diagram (Plate II.).

STRUVE during his observations discovered a star (II.) which, strange to say, did not occur in HERSCHEL's list, although it was brighter than several small stars in this neighbourhood which were given in that list; nine days afterwards he saw it with difficulty, while the other stars had preserved their brightness.

He repeatedly compared this star with 88, 51, 57, 75, 78, with the view of determining, if possible, the period of their variability, and gives the maximum and minimum brightness as follows, 88, whose brilliancy he found to be most constant, being taken as of the 12th magnitude:—

51 varies from 11·9 to 12·5 magnitude.
57   „   „   12·5 to 13·5
II.   „   „   11·8 to invisibility.
III.   „   „   12·5 to   „
75   „   „   12·0 to   „
78   „   „   12·5 to   „

98 he also considers variable to a certain extent; and of the above list, II. he thinks is a variable star, whose maximum is of short duration and minimum of long duration, and 75 one whose maximum is of a long and whose minimum is of short duration.

Another instance of variability is IV., which STRUVE considers varies from the 11·5 to 13·5 magnitude.

With reference to the variability of the stars, our observations furnish very little information; our attention was directed principally to delineating, as carefully as possible, the various details of the nebulæ with reference to stars whose positions had been previously determined. I may, however, remark that 29, to which STRUVE has assigned the same ($7\frac{1}{2}$) magnitude as that of 20, 23, and 24, is now decidedly fainter than they are; it is about the 11th or 12th magnitude. HERSCHEL, on the other hand, gives the following magnitudes, 20=23=24=8th magnitude, 29=12th, which agrees with their present appearance; 29 also is not quite in the same relative position as given in HERSCHEL's drawing; he represents it nearly in the same line with 20, 23, 24, whereas the line through 23, 24 is inclined through an angle of about 28 degrees from the line 24, 29. This discrepancy is probably due to an error in the position of 29, as HERSCHEL places this star in his fourth or least accurately placed class. The discrepancy seems far too great to be accounted for in any other way.

2. *Variability of Form and Intensity of the Nebulosity.*—On this subject it is impossible to speak decidedly.

On comparing the following six drawings,—

Sir J. HERSCHEL's of about the year 1825,
Sir J. HERSCHEL's   „   „   1837,
Mr. BOND's   „   „   1848,
M. LIAPOUNOV's   „   „   1850,
Mr. LASSELL's   „   „   1854,
Mr. HUNTER's   „   „   1863,

great discrepancies exist in almost every part, but these are probably to be attributed in a great measure to the difference of power in the instruments used and the amount of labour expended on the drawings, as no continuous change seems to be shown by them. In the case of the spiral nebula round 108, BOND's, LASSELL's, and HUNTER's drawings appear to agree tolerably well, allowance being made for the difference of size of the instruments, but when we go back to HERSCHEL's drawing of 1837 we find a considerable discrepancy. HERSCHEL's drawing of 1825, however, as far as it goes, is in this place more like the latter drawings. With regard to the following extremity of the Huygenian region, all the former drawings, with the exception of LIAPOUNOV's, represent the "Frons" as curving round to meet the proboscis major, which latter also curves round to meet the former; whereas Mr. HUNTER represents both these parts as curving slightly in the opposite direction. This I am satisfied is their present appearance. If, however, the night is not good, they acquire very much the appearance of the other drawings, the light of the brighter portions being scattered, to a certain extent, over the intervening space. With regard to the shape of the preceding edge of the proboscis, the drawings of HERSCHEL (1825), BOND, and LIAPOUNOV represent it as of a uniform curve throughout the greater part of its length, one elbow only at star 126 being shown in those drawings, which extend far enough south; whereas in the case of the other drawings, more powerful means bring out more irregularity of outline. In the case of the Huygenian region, HERSCHEL's drawing (1837) agrees much more nearly with Mr. HUNTER's than any of the others, although the interval (30 years) is so much longer than in the case of Mr. BOND's and Mr. LASSELL's drawings (15 and 9 years respectively).

With reference to the relative brightness of the various parts, I find recorded by Mr. HUNTER, Feb. 22nd, 1861, " In bright moonlight the degrees of brightness are—

" 1. The Huygenian region.
" 2. The nebulosity immediately south preceding it.
" 3. The Mairian region.
" 4. The subnebulous region.
" 5. The south Messierian branch, and the nebulosity immediately north of the Huygenian region."

And again, " The Observation of February 22nd, 1861, gives very different degrees of brightness for the various regions from what they had this season (1863–64).

" 1. The Huygenian region.
" 2. The nebulosity immediately south preceding it.
" 3. The nebulosity immediately north of it.
" 4. Subnebulous region.
" 5. The south Messierian branch and the Mairian region nearly equal."

Mr. HUNTER on two occasions estimated, as nearly as he could, the relative brightness of the various masses of nebulosity of the Huygenian region. The following are his estimations (see diagram)*.

* To these estimates we may attach much importance, as Mr. HUNTER had the advantage of a considerable amount of training as an artist.

o

February 13th, 1864.

$\sigma, \tau, \nu, \gamma$ nearly equal; brightest of
    these is perhaps σ.

π.

$\omega, \epsilon, \delta, \beta$; β is the faintest of these four.

$\alpha; \iota, \psi, \lambda$.

ζ, φ, faintest.

March 1st, 1864.

σ, brightest.

ν, τ.

γ, ϵ, δ.

θ, μ.

π, very faint.

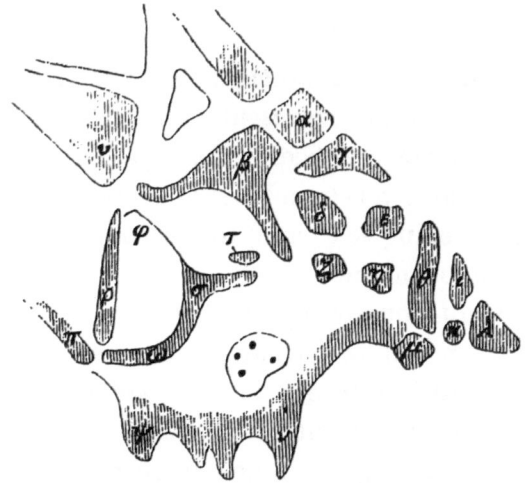

There are several places where we have reason to suspect that a change of form may have taken place in the nebulosity since our observations commenced. 1st. In Mr. BINDON STONEY's drawing, of which an outline is given at the upper right-hand corner of the Skeleton Map, a dark lane exists running from 88 in a direction parallel to the "Frons," whereas at present the only break in the nebulosity at all in the same direction runs from 88 in a south following direction.

2nd. The projection of the nebulosity below 88 into the Sinus Magnus does not exist in Mr. STONEY's drawing.

3rd. The following outline of the nebulosity immediately below 75 is concave towards the following side in Mr. STONEY's drawing, but convex in Mr. HUNTER's; in all these points I believe that Mr. HUNTER gives as nearly as possible the present appearance.

4th. Mr. HUNTER represents the outline of the nebulosity surrounding the dark region or lake round the stars 32, 35 as very marked; I often examined this part during the seasons 1864–65 and 1865–66, but never saw it quite as distinctly as it is represented on the following side, nor did I see the elbow just following 35 ; the nebulosity appeared to be more of the shape represented by the coarsely dotted line in the Skeleton Map.

5th. I was never able to see more than two of the three rays below this lake, and except on two or three occasions I could only make out one.   Mr. HUNTER has since told me that in the last season during which he was working, these rays were much fainter than they had been previously, and that they are represented too bright for their appearance during the season 1863–64.

In connexion with this subject it may not be uninteresting to compare the observations of former observers with each other and with our own.

Sir J. HERSCHEL in his paper of 1825 discusses the differences between his own drawings and those of HUYGENS, PICARD, MESSIER, and LE GENTIL, and thinks that the first three, when compared with his own, tend to show a gradual diminution or condensation of the nebulosity; but LE GENTIL's, which was older than MESSIER's, represents it just as he himself saw it.

We next come to Sir J. HERSCHEL's paper of 1837, in which he says that although to any one who has not viewed this object through powerful telescopes the differences between the various drawings, including his own of 1824 and 1837, may seem great, and tend to convey a strong impression of great and rapid changes undergone by the nebula itself, yet, after carefully comparing his own two drawings, he comes to the conclusion that the differences are not greater than he is disposed to attribute to his own inexperience in such delineations in 1824, to the greater care bestowed on the later drawing, and especially to the advantage of better local situation and superior defining power, &c. of the telescope at the latter date (Cape Observations, page 31).   There are three points, however, to which he directs attention, but in the case of two only of them is he inclined to conclude that there is any evidence of change; these points are—

1. The form and position of the nebula oblongata between 127_{1} and 129_{1}.
2. The position of the nebulous spur between 111 and 122.
3. The form of the nebula round 108.

In 1824 Sir J. HERSCHEL saw the nebula oblongata as a "tolerably regular oval," nearly in a line between the stars 120 and 136, whereas in his drawing of 1837 it is irregular in outline, and decidedly above the line through 120 and 136.

With respect to the form of the nebula oblongata, the brighter part forms a "tolerably regular oval," but when the fainter parts are included it seems to be more of the form given in HERSCHEL's drawing of 1837.   It is therefore quite possible, even probable, that HERSCHEL would have seen it oval in 1825, but long and slightly curved upwards with the superior means at his disposal in 1837, without any change of form having taken place in the interval; but as regards its position it appears to be now entirely above the line 120–136.

With regard to the nebulous spur between 111 and 122, diagrams which he made in 1832 and 1834 represent it as "running directly from 135 to 111 and forming a complete hook no way disjoined from the proboscis."   In 1837 he saw it "neither joined to the proboscis nor directed towards 135, but rather towards a point one-third the distance from 135 to 126" near the position of 131.   HERSCHEL's second drawing appears to agree very fairly with the accompanying one in this respect; perhaps the superior definition of HERSCHEL's instrument in 1837, a better atmosphere, and the greater * meridian altitude of the object enabled HERSCHEL to perceive the interval between this spur and the proboscis which had escaped his notice in 1832 and 1834.

* This last applies to the diagram of 1832 only.

With regard to the nebula round 108, the amount of detail in HERSCHEL's drawing of 1837 is so much greater than in that of 1824, and the detail in the accompanying drawing is so much greater than in HERSCHEL's of 1837, that it seems hardly possible to arrive at any conclusion by comparing them.

Resolvability.

On this subject HERSCHEL remarks in his paper of 1825 that the illumination of the Huygenian region is " extremely unequal and irregular," and compares it to " a curdling liquid, or a surface strewed over with flocks of wool, or to the breaking up of a mackerel sky when the clouds of which it consists begin to assume a cirrous appearance; not very unlike the mottling of the sun's disk, only the grain is much coarser and the intervals darker, and the flocculi instead of being generally round are drawn out into little wisps. They present, however, no appearance of being composed of small stars, and their aspect is altogether different from that of resolvable nebula." This describes very well the appearance on any very moderately good night, but at this time HERSCHEL does not appear to have seen clearly the coarser mottling or breaking up into nebulous masses which have since been seen and drawn. This curdled appearance he does not find described in any previous account.

The next mention of this subject appears to be by Mr. BOND, who, after quoting part of HERSCHEL's remarks given above, says, " To me it appears composed of several clusters of stars, the components being separately seen for a moment under favourable circumstances, more particularly north of 75 *, and likewise in the vicinity of 8 and 91; but where the nebula assumes a cirrous character, as in the Messierian branch, I can see nothing of the kind."

The next mention of the subject is by Mr. LASSELL, who says that with a power of 1018 the whole nebula " seemed like large masses of cotton wool packed one behind the other, the edges pulled out so as to be very filmy;" also " the brightness of the minute points about the trapezium is strikingly greater than at Starfield, yet I could not mark the places of more than three or four new stars." " With a power of 1018 there is no appearance of resolvability." " With a power of 1018 the wool-like masses appear as I have previously described them, and there is no disposition whatever in them to turn into stars," &c.

We now come to our own observations made with the 3-feet and 6-feet telescopes, and find the following remarks:—

February 17th, 1849.—" Saw a multitude of stars in the bright part about the trapezium, but when they came to be drawn, only got in 9 certain and 6 uncertain, the state of the air having become worse."

January 22nd, 1851.—" At times the 5th and 6th stars of the trapezium seen, also red stars south following the trapezium," &c.

Mr. HUNTER on many occasions saw small stars in the brighter parts; the following are nearly all his remarks on this subject:—

* N.B. In all cases where not otherwise specified the numbers are those of HERSCHEL's list.

December 4th, 1861.—" I thought I could resolve the Huygenian region at ν, ζ, and ι."

February 22nd, 1862.—" 3-feet speculum newly polished shows it much better than the former, indeed better than the 6-feet now in the tube. The part round the trapezium looks just like fine flour scattered over a grey surface, so that I have no hesitation in saying that it is composed of stars, many small ones seen through it," &c.

January 12th, 1864.—" The knots γ, δ, ϵ, ν are, I think, resolved "*.

February 3rd, 1864.—" $\varkappa$ is decidedly resolvable, ι is also resolvable ; I believe about θ is also resolvable "*.

February 4th, 1864.—" At moments I could see stars through the Huygenian region glancing, and I have no doubt that all the bright knots of it are resolvable ; at $\varkappa$, δ, ϵ, γ I saw clearly at least one star in each, and at δ I believe for a short time I saw its stars separated." The little knots opposite τ and σ have each a resolvable look.

March 1st, 1864.—" The Huygenian region is clearly resolved ; I could see the individual stars, though I could not count them ; the stars are well separated in the triangular knot α. I strongly suspect the region at 113_{m} is resolved, especially at the edges of the bay."

March 10th, 1864.—" Saw stars clearly at σ and at ν*, at intervals also around 113_{m}."

March 24th, 1864.—" Huygenian region clearly resolved."

From these observations we may conclude that there are multitudes of small stars in the whole of the bright parts round the trapezium, and also at 113_{m}. Mr. HUNTER makes no mention of resolvability in the proboscis major, but he has marked the preceding edge from 126 to about 80″ below this star as resolvable. All the parts which have this appearance are marked over in the drawing with dots of Indian ink, and the rest of the nebula was done with a stump and blacklead pencil. It was, however, found almost impossible to reproduce this difference of appearance in the engraving, since the whole of the surface consists of minute black dots.

This resolvable appearance can be seen on good nights only, and with a very good speculum.

We now come to the last part of our subject, the knowledge of the nature of this nebula acquired by the use of

The Spectroscope.

Very little has been done with this instrument as yet. The clock-movement for the six-feet telescope is not yet finished, and consequently it is impossible to keep the slit steadily on any small object ; we have, however, obtained the spectra of about twelve objects, and our results, as far as they go, fully confirm those of Mr. HUGGINS, namely, that some nebulæ give gaseous spectra, consisting of one, two, or three bright lines, while others give faint continuous spectra in their brighter parts. The telescope which was generally used was the three-feet, in the Herschelian form ; of course without clock-movement. This instrument was found on the whole more convenient than the six-feet, as it could be kept with greater ease on the object.

The following was the method of observation employed. One observer kept the ob-

* See diagram, p. 67.

ject as well as he could on the slit by viewing with a lens its image in a diagonal reflector placed in front of the slit, and having its edge almost touching the slit, but not actually covering it, and swinging the tube till the brightest part of the nebula just passed the edge of the reflector, while the other observer looked into the spectroscope. The telescope was used in the Herschelian form, as it was desirable to get as bright an image as possible, but not of importance to get the best possible definition.

The following is a list of the objects which we have examined, some of which have also been observed by Mr. HUGGINS. All these objects were examined both by Mr. BALL the present Assistant, and myself alternately.

Gaseous Spectrum.	Continuous Spectrum.
Great nebula in Orion.	Great nebula in Andromeda.
2102	2373
2343	2377
4964	2786
	2203
	2207 No decided spectrum seen;
	2211 spectrum suspected to be
	2347 continuous.
	2786

Although the last six gave no decided spectrum, there can be very little doubt but that their spectra are continuous; they were examined before our eyes had been accustomed to a faint or continuous spectrum by examining those of brighter objects, such as the nebula in Andromeda and the bright cluster in Canes Venatici. If they had given a gaseous spectrum we could hardly have failed to have seen it.

Although in consequence of the smallness of the number of objects hitherto observed it would be premature to lay much stress on any inferences derived from these observations, it may not perhaps be out of place to mention that in addition to the results arrived at by comparing our observations with those of Mr. HUGGINS (viz. that no cluster or resolved nebula yet observed gives a gaseous spectrum, and of the remainder those which give a continuous spectrum are generally of a more resolvable character than those which give a gaseous spectrum*) we find—

1st. That no planetary or annular nebula yet observed has been found to give any but a gaseous spectrum, with in some cases a suspicion of a very faint continuous spectrum.

2nd. That no nebula of the class generally denominated "rays" has yet been found to give a gaseous spectrum.

3rd. That of the remaining objects observed, those which give a gaseous spectrum but are not denominated planetary are in one respect of the same character with them, viz. that they have in many places a well-defined termination to an almost uniformly bright

* See Mr. HUGGINS's paper, Philosophical Transactions, Part I. 1866.

nebulosity, whereas those nebulæ which give a continuous spectrum appear to fade away tolerably uniformly on all sides from their nuclei, and although in some cases they have dark lanes running through them, the edges do not generally appear very well defined.

Spectrum of Nebula in Orion.

With regard to the spectrum of the nebula in Orion, three bright lines were seen several times, both with 3-feet and 6-feet instruments, but no attempt was made to identify them except on one occasion, when the spectrum of the nebula was compared with that of the electric spark in a capillary vacuum-tube containing a trace of hydrogen, similar to those used by M. Plücker in his researches on the spectra of gases, and one glimpse of the coincidence of the green line in the latter with the most refrangible of the three lines in the former was obtained. The least refrangible line was the brightest, the most refrangible was next in brightness, and the middle line the faintest. Both Mr. Ball [*] and I were almost certain that there was, in addition to the three bright lines, a faint continuous spectrum; to me there appeared to be a dark space on the less refrangible side of the least refrangible line, of breadth about equal to the distance from the same line to the second, and beyond this a faint light dying gradually away towards the red extremity. A continuous spectrum would probably explain this appearance, as the intensity of the bluish-green and green on the more refrangible side of (b) in a continuous spectrum is much less than that of the yellowish-green and yellow. We also once suspected a very feeble light at the other side of the three bright lines.

It might at first sight appear that these observations and those of Mr. Huggins on the spectrum of this nebula lead us to results which are completely at variance with those derived from our numerous observations, and those of Mr. Bond on the resolvable appearance of the Huygenian region and other parts of the nebula; but when we consider the subject carefully we shall see that this is far from being the case.

Reason why no continuous Spectrum was seen.

It is evident that when a spectroscope whose collimator and telescope have object-glasses of equal focal length is placed with its slit in the plane of the image of a nebula giving out perfectly homogeneous light, a line of light of length and breadth equal to the length and breadth of the slit will be found at the focus of the spectroscope-telescope, and of brightness equal to the brightness of the part of the original image on the slit at the time, multiplied by a constant quantity (E) less than unity, depending on the number of reflecting surfaces in the apparatus, &c.; and if the nebula give out light of three different refrangibilities, the mean brightness of the lines will be one-third of this quantity; whereas if the light emitted be of all refrangibilities the mean brightness will be less in the ratio of the length of the spectrum to once or three times the breadth of the slit, according as we compare it with a spectrum of one or three lines. This is fully confirmed in practice; for in the case of those nebulæ whose spectra are continuous, the

* The present Assistant.

spectrum is extremely faint, except just at the nucleus, and even there it is pretty faint. When therefore we consider that a great part of the light of the Huygenian region of the nebula in Orion, the brightest parts of which are probably not much, if at all brighter than the nuclei of such nebulæ as 2373, 2377, goes to form a gaseous spectrum, we can hardly expect that the remaining light could produce any but the feeblest continuous spectrum. As a further confirmation of this view, it may be mentioned that on one occasion the spectrum of this nebula was examined in bright moonlight; the light was so strong that the Huygenian region was scarcely visible; in fact its boundaries were not more apparent than the boundary of the proboscis major at 131 is on a dark night; yet although the three gaseous lines of the nebula were very fairly seen, no continuous spectrum from the moonlight, which was probably equal in intensity to the light of the nebula, could be detected.

This observation, however, was not so satisfactory as I could have wished, as a haze which soon after eclipsed the nebula was beginning to come on at the time.

Besides the central parts of the nebula, the nebula Mairiani was examined with the spectroscope, and a gaseous spectrum seen near the star 108. The proboscis major was also examined near 126, and a gaseous spectrum seen. Orionis was also examined, but no spectrum but that of the star itself detected.

A clock-movement for the 6-feet telescope is now in progress, and when this is completed we hope to examine the spectra of this and other nebulæ with more care.

MEMORANDUM.

Through the kindness of Sir JOHN HERSCHEL I have been permitted to see his remarks on this paper, and gladly take advantage of his suggestions. The engraving is upon the whole very accurate; a little more softening off in the faint outlying parts would have been desirable, but Mr. BASIRE did not think that it would be practicable consistent with the reasonable durability of the plate; the forms, however, are correct. The sharpness of outline and the hard and marked character of the principal features are the result of the great light of the instrument; with a diminishing aperture these characteristics gradually fade away. The engraving faithfully represents the object as it may be seen on any clear night, and the details are so well marked that no material change can take place hereafter which will not at once be recognized with an instrument of similar power. The interior of the trapezium has not been examined recently with the view to the question whether it is absolutely dark. With the 6-feet instrument the eye is so dazzled by the light of the four stars that it is difficult to form an accurate opinion; and any nebulosity which may exist is probably too faint to affect the spectroscope. I am not certain that any part of the nebula is absolutely free from nebulosity, but the contrast is so great between the dark spaces alluded to by Sir JOHN HERSCHEL, and the contiguous portions of the nebula, that even in the drawing it was scarcely possible to indicate nebulosity so slightly as not to interfere with the proper gradation of light; in fact it was scarcely possible to represent the bright parts sufficiently bright.

MDCCCLXVIII.

```
╔══════════════════════════════════════════════════════════════╗
║                                                                ║
║        A CONTRIBUTION TO THE HISTORY OF IRONCLADS              ║
║              from the Correspondence of the late               ║
║                 Earl of Rosse, 1854-1865                       ║
║          Published by the Institution of Naval Architects      ║
║                                                                ║
╚══════════════════════════════════════════════════════════════╝
```

INTRODUCTORY NOTE. BY THE RIGHT HON. THE EARL OF ROSSE, K.P., D.C.L., LL.D., F.R.S.

FINDING in the Journal of the Society of Arts, in the report of Sir William White's lectures of last year, the statements that "the credit of initiating the construction of the earliest completed ironclads belongs to the Emperor Napoleon III," and, again, that "these ironclads were floating batteries, built in 1854, and used during the Crimean War," it struck me that it might be of interest to look up old correspondence bearing upon the subject, which I felt sure would somewhat modify those statements.

I knew that my father had done what he could in conversation and by letter to urge our Naval and Military Authorities to take up the construction of armoured vessels for the attack on the Russian forts, and I understood that the part the Emperor Napoleon III took was, on learning of the proposal, to press the matter forward and get our authorities to take it in hand at once, when they were hesitating to move. Floating batteries were then put on the stocks, but were not, I believe, completed before the peace.

Were it not that the ancient history of ironclads has been brought into notice by being made the subject of the important course of lectures above referred to, I should have hesitated to put forward letters which, after the great advance of the applications of mechanical engineering, to Naval purposes during the last half century, would seem quite out of date.

I remember the tracing referred to in Letter VIII (February, 1855). My father thought the proportion of beam to length insufficient for combining the necessary resistance to penetration with buoyancy. He was also disappointed that the main structure of the hull was to be of wood.

The Howitzers referred to in Letter VII (December, 1854) are still in my laboratory, and a punt-like model, 3 ft. 3 in. by 10 in. by 4 in., probably represents my father's general ideas as to proportional dimensions.

P

The letters have all been copied from my father's press-copy book, except the last two, which are from Wrottesley's "Life and Letters of Sir John Burgoyne," there being no copies in my father's book.

ROSSE.

Birr Castle, February 16, 1907.

LETTER I.

TO SIR HOWARD DOUGLAS, BART., Author of "A Treatise on Naval Gunnery."

13, *Connaught Place, June* 11, 1854.

DEAR SIR HOWARD,—I return the pamphlet with many thanks, having read it with much interest. The principle laid down that assailable points should be guarded by forts, so as to leave the fleets free, appears to me to be but common sense. The Czar has acted upon it ; we, I think, are not yet alive to it. As to the experiment relative to the resistance of plate iron, the thickness of the boiler plate is not given, nor the size of Commodore Stockton's gun. Boiler plate usually runs from $\frac{3}{8}$ in. to $\frac{5}{8}$ in. thick ; taking it at $\frac{1}{2}$ in., the largest would have been $2\frac{1}{2}$ in thick, while 5 in. seems the minimum for 32 lb., so that if we add 2 in. as a reasonable margin, the largest should have been 7 in. thick, which, probably, would have withstood Commodore Stockton's gun. Even if the iron was to be 12 in., a properly proportioned vessel would carry the weight, even though it was not immensely large.

Believe me to be,
Dear Sir Howard,
Truly yours,
ROSSE.

LETTER II.

Letter from the EARL OF ROSSE to Sir JOHN BURGOYNE.

13, *Connaught Place, June* 12, 1854.

DEAR SIR JOHN,—You are probably worried with conundrums for destroying the Russian ships under their forts ; this is mine. To build an iron steamer, proof against shot, shells, and boarders, and to run at the enemy's ship and sink it with one blow of the cutwater. To construct such a steamer I have no doubt is perfectly practicable ; and I do not think it would be a very long business with the resources our great shipbuilders have at command. The data we have would be sufficient to enable us to calculate the lines of the steamer, allowing, however, a very large margin of excess of strength : to prevent waste, a more accurate determination of the data would be necessary. First, the requisite thickness of iron. The best data I know of are the French experiments, detailed in the third edition of "Sir Howard Douglas on Naval Gunnery." From these experiments, I think we may assume that 3-in. iron plate would be shot-proof, except at a very short range, and where the impact was perpendicular ; 4-in. plate, I think, would be safe under all circumstances which could occur in practice. The steamer should have no bulwarks, and when trimmed, the deck must not be more than 12 or 14 in. above the surface of the water. The first 2 ft. of the side would be 4 in. thick ; the next foot, being considerably protected by the water, would be 3 in. ; and the fourth foot 2 in. ; the remainder of the hull ¾ or perhaps 1 in. thick. The draught of water would depend upon the size and shape, but with the machinery, and *without the deck*, it need not, I think, exceed 8 ft. As to the deck, I think it has been found that two layers of balk beams 12 in. thick each, crossed, is proof against heavy shells ; I am speaking however, only from a very distant recollection of the facts given either in the "Aide Mémoire," "Jones' Sieges," or the "Professional Papers." I suppose that to be so ; then 2-in. iron plate would be about an equivalent ; and as iron is nearly eight times as heavy as water, a deck of 2-in. iron plate would increase the draught of water 16 in., so that, allowing for deck beams and fastenings, the total draught might be, perhaps, 10 ft. Before the lines of such a steamer were calculated, the best data should be obtained.

(1) I think it would be desirable to repeat the French experiments, employing a succession of plates riveted together, which is the way the required thickness would probably be obtained in practice. (2) To obtain the best information as to the impact of heavy shells. (3) The amount of protection which might be relied upon from the water at different depths. Should it be found that greater strength would be required, either in the deck or sides, it would only be necessary to increase the size : the great principle that the buoyancy increases much more rapidly than the surface would make all difficulty surmountable. The funnel need not appear above deck, as a fan would answer instead, and the hatchway could be effectually secured against boarders, the vessel being steered below. Small holes for muskets would probably be sufficient to keep the deck clear and prevent grappling.

Ships in docks would of course be safe, but I think nowhere else where there was a fleet outside. You may, perhaps, discover some fatal objection to all this, but if not, perhaps it might be worth considering.

Believe me, etc.,

ROSSE.

In the rough calculation I have assumed 300 as the nominal horse-power.

LETTER III.

TO SIR JOHN BURGOYNE, BART.

13, *Connaught Place, June 26, 1854.*

DEAR SIR JOHN,—As a little supplement to my former hints, I will add that I have roughly examined the question whether it would be practicable to construct a vessel proof against shot and shells capable of carrying an efficient breaching battery, and yet not so large as to make the expense a fatal objection. The result is that I think the project quite feasible. As such a vessel would probably be employed in water nearly smooth, I think she need not carry her guns more than 3 ft. above the water. The exposed sides of the vessel would be of plate iron 5 in. thick, and her ports might be round and a little larger in diameter than the guns. Her deck would be about 6 ft. above the water, and 2 in. thick of plate iron. No bulwarks. Probably you would consider sixteen heavy guns sufficient, and that eight guns at fifty yards would in a reasonable time effect a breach in one of the Cronstadt forts. Such a vessel need not draw more than 12 or 13 ft., and, if the little charts in the booksellers' shops are trustworthy, there is more than that depth of water close to the forts. About 1,500 tons would, I think, be large enough, with a beam excessive in proportion to her length, to give the necessary buoyancy and stability under top weight. All this is the roughest possible, but I think if worked out in detail the result would not be widely different. The greatest care would, of course, be necessary to guard against submarine explosives. Should the Russians cast a few enormous guns to meet the attack of such a ship, I should think, by a little management, it would be possible to take up a position out of the line of their fire, and while taking up that position so to steer as to prevent the ship from being struck, except at a very oblique angle.

I have quietly spoken to several of our fellows, three of them civil engineers, and one an artillery officer, and I have not heard any doubt expressed as to the practicability of effecting the object in the way I have suggested.

Believe me to be,

Dear Sir John,

Yours truly,

ROSSE.

LETTER IV

From SIR JOHN BURGOYNE, BART.

Ordnance Office, Pall Mall, July 2, 1854.

MY DEAR LORD ROSSE,—I have many apologies to make for not having answered your two letters earlier, but as they are on very interesting subjects, I was in hopes before I replied to them to have obtained more information than I possess on the matter, and that, after all, I have not been enabled to do, but prefer writing in an unsatisfactory manner rather than appear to be guilty of neglect.

I presume that the application which your Lordship proposes for your naval construction of vessels is rather to show the value of the idea should it prove to be well founded—than with the intention of entering into much discussion about the particular *modus operandi;* indeed, as the whole depends upon the principle of the shot and shell proof vessel, the real question is to establish the practicability of that principle in the first instance, and ascertain the data necessary for that object before drawing on discussion as to the manner of applying it.

If one could obtain shot and shell proof surfaces by any at all moderate degree of substance, weight and expense, it might be turned to very useful account by land and sea in various ways. I do not know on what ground you conceive that iron of 5 in. thick (*sic*) would answer, but my impression is that against a 32-pounder at a high velocity and vertical impact, however manufactured, it would be thoroughly shattered by even a single shot supposing the surface to be considerable, such as the side of a ship without cross supports within. Iron is very treacherous, and breaks, rends and tears under very irregular efforts. The Navy have a thorough dislike to it for the sides of ships, but then they have never contemplated, I believe, such thickness. There is then the question of the effect when struck at different angles. By trials against massive stone walls, the shots glanced off to the wall being at an angle of about 77 deg., after which the shot glancing has a tendency to penetrate, but it is a curious circumstance that in every instance of the shot glancing off the wall, it (the shot) invariably broke in two or more pieces. The shot, however, was cast, whereas your walls whould be, of course, malleable or rolled. One matter of experience in favour of your system would be that a ship cannot be struck with any force by a shot more than 2 ft. under water, if so much. The only trials of shot-proof sides for vessels that I am aware of was for the floating batteries to breach Gibraltar in 1782. D'Arçon, the French engineer who invented them, wrote an apology for their want of success : it is a very rare work, I have never been able to see it, but I understand that he had a close network of iron, which was some feet deep, and the square chequered interstices which were not above 1 in. or 2 in. square, filled with hard wood. He said that, by experiments at Cadiz, it was proved to be shot proof, and that, in the end, they were not set on fire by the red hot shot, but by their crews. With respect to your decks, plunging shot and shells may hit them very hard—but if two or three men of such knowledge as your Lordship, and those you mention, conceive the success even *probable*, most decidedly it ought to be tried, but first at a target before putting it in the form of a ship.

Yours very truly,

JOHN BURGOYNE.

LETTER V.

TO SIR JOHN BURGOYNE, BART.

Athenæum, July 3, 1854.

DEAR SIR JOHN,—The grounds upon which I took 5 in. as sufficient were the French experiments at Metz, given in the third edition of Sir Howard Douglas' "Naval Gunnery," page 164. It is there stated that a 24-pounder with a charge of 4 lb. 4 oz. at 66 ft. distance did not pierce a plate of iron 3·08 in. thick ; with 6 lb. it did so. From this some estimate may be made, thought a very rough one, of the thickness necessary to resist a 32-lb. shot with a full charge, and it accords pretty well with deductions from theory. It appears that iron $\frac{5}{8}$ in. thick almost breaks the shot to pieces (page 167), so that the shot does not act upon very thick iron plates by virtue of its form, but mainly by its momentum. The power of resistance of iron plates under such circumstances would increase much faster than their thickness. This kind of reasoning is no doubt very unsatisfactory in comparison with experiments, and the only inference which at present I should venture to draw is that there is a certain thickness of plate iron which would resist 32-lb. shot, and that a ship of no very great size would be able to carry it, provided it was properly shaped and very low in the water. At one of the Royal Society soirées a 9 or 12 lb. shot was exhibited, which had been laid upon a block of Indian rubber and struck with Nasmyth's hammer. The shot was partially flattened and fissured all round the edge, just as if on the point of going to pieces ; the Indian rubber was uninjured. Here we see the cohesion of the shot was not sufficient to bear the strain of being forced into the Indian rubber ; and the shot been projected, even with a small velocity, its *momentum* no doubt would have carried it through.

Believe me to be,

Dear Sir John,

Very truly yours,

ROSSE.

LETTER VI.

TO THE DUKE OF NEWCASTLE.

July 6, 1854.

Lord Rosse presents his compliments to the Duke of Newcastle, and begs to say that he has been considering, no doubt in common with many others, in what way the great mechanical resources of England could be brought to bear against the mechanical defences of St. Petersborg. The records of sieges show beyond question that a few heavy guns within a range of 50 or 100 yards will soon demolish the strongest mason work. The question therefore is, could a floating battery be constructed proof against shot and shells and yet sufficiently buoyant to carry a few heavy guns with the necessary supply of ammunition ? This appears to be quite practicable. First as to the material, the Spaniards tried wood against Gibraltar and failed. Their floating batteries were destroyed by red hot shot that now, although by modern processes wood can be rendered in some measure fireproof, still modern gunnery would destroy wooden batteries however strong, as shells fired horizontally would lodge, and bursting would soon blow them to pieces. Plate iron appears to be the suitable material. The next question then is how thick the plate iron should be. No doubt there is a certain thickness which would be quite effectual. Strange as it may appear, no experiments seem ever to have been tried either by our naval or our military authorities on the resistance of plate iron to artillery. Recently, indeed, experiments have been tried at Portsmouth on the splintering effects of shot passing through thin plate iron. These experiments were tried *after* several war steamers had been built of thin plate iron, and the result was the steamers were condemned for war purposes. Strange to say the authorities were then satisfied and proceeded no further with the experiments, instead of completing the inquiry by investigating the other case, the effect of shot on thick plate iron of sufficient substance to stop the shot altogether. The only experiments of the kind have been made at Metz, and the results are given in the third edition of Sir Howard Douglas' "Naval Gunnery," but they were made with a different object, and are quite insufficient as data for the purpose of computing the tonnage and proportions of iron floating batteries. As far as can be inferred from the French experiments (but it would be unsafe to trust to theoretical deductions from such imperfect data), the minimum thickness of plate iron, or rather a series of plates of iron, to resist the direct fire of 32-lb. shot should be 5 in., and to resist the descent of heavy shells it should not be less than 2 in. Upon these data Lord Rosse computed in the roughest manner the proportions of two screw steamers, one *without* armament but with considerable steam power which might enter a harbour however guarded by batteries, and running full speed at any ship at anchor might sink it with one blow of its cutwater ; the other, a steamer with small steam power but with a few heavy guns under a bomb-proof deck to be laid as close as possible to the fort to be breached. He sent the results of these very rough computations to Sir John Burgoyne about a month ago. Sir John Burgoyne very properly observes that till the effects of shot upon plate iron shall have been better ascertained, it would be impossible to form any accurate opinion, but that the subject is one of much importance. Should the Government think it advisable to direct the authorities at Portsmouth to make the necessary experiments, a target could be put together in a few days, and the preliminary experiments might be computed in a week. How far the experiments should be carried would depend very much upon the first trials. The data having been accurately ascertained any of our leading engineers would have no difficulty in computing the size, proportions, and probable cost of a shot-proof steamer suited to the required purpose. Lord Rosse has mentioned this matter to a few of the Fellows of the Royal Society best acquainted with such subjects, and no serious difficulty has been suggested ; he has refrained, however, from mentioning it publicly.

LETTER VII.

TO SIR BALDWIN WALKER.

Castle, Parsonstown, December 16, 1854.

DEAR SIR BALDWIN,—I am sure you will think it rather impertinent in a landsman to give you a hint, but I cannot refrain from doing so. Last spring, having ascertained by computation on ordinary mechanical principles within small limits of error the thickness of plate iron which would resist 32-lb. shot under various circumstances, a very easy matter with the excellent modern engineering data we have ; and having in some degree verified the results by comparing them with deductions (?) from the French experiments, I computed roughly the proportions of shot-proof ships, which with the necessary strength would have the necessary buoyancy, and I sent the result to Sir John Burgoyne. A difficulty stared me in the face *in limine*, that such vessels though without masts might be liable to capsize in a heavy sea as there was so much top weight ; but a little simple calculation made it evident that the required stability could be obtained without sacrificing essentials, and I did not think it necessary to allude to that point in writing to Sir John Burgoyne. It now occurs to me *barely possible* that in constructing your shot-proof ships there may be the oversight of omitting to make the necessary provision for stability in a heavy sea, in which case there might be a serious disaster. The required calculations are perfectly simple when the data are determined, but simple things are sometimes overlooked, and it is quite possible that a ship apparently stable in smooth water might in a heavy sea turn right over. I have therefore ventured to throw out this hint, as perhaps you might think it expedient to mention the subject to your naval architect.

There is another point. In writing to Sir John Burgoyne I assumed that the projectiles to be resisted were 32-lb. spherical shot, and on that assumption I obtained 5 in. as the minimum thickness at a perpendicular incidence and at very short ranges. At that time (the first week in June) heavy rifled ordnance had not, as far as I was aware of, been employed with much success. Now it seems to be otherwise, and the ships, or rather floating batteries, may find at Cronstadt antagonists far more formidable than the ordinary 32-pounder. Of Mr. Lancaster's rifled ordnance I know nothing practically ; I have never seen them. The contrivance appears to me to be very unmechanical, and I should not be surprised if the Czar has something very much better in spring. The scientific Russians I have occasionally been thrown in contact with have always represented the Czar as a very able man thoroughly conversant with everything relating to the *matériel* of war. He has a magnificent foundry with every mechanical appliance. He has Jacobi and other ingenious and able men at his elbow. It is very probable they will suggest to him that to sink your floating batteries as described in the newspapers, it is only necessary to cast heavy rifled ordnance. This he can do with great facility. In the present advanced state of mechanical engineering guns can be cast hollow, sound, and smooth enough for immediate use. Woolwich people would perhaps smile at this, but it is nevertheless the fact. Four or five years ago, requiring a few iron guns for some experiments, they were cast in my laboratory : we cast them hollow, and they might have been at once grooved ; but, as I wished the bore to be quite smooth, a cutter was passed through. The surface of the bore was free from specks, and the iron as sound as if the guns had been cast solid : the largest was but an 18-lb. Howitzer gun, but there was nothing to prevent their being cast of any size. The cylinders of the largest steam engines are every day cast in the foundries of England and Scotland perfectly sound and free from specks. Instead of the core being round, it might have had the oval twist of Lancaster ; or, what I should have thought much better, it might have been formed so that the gun should have had two *rectangular* broad grooves to receive long *rectangular* prominences on the sides of the elongated projectile. There would be no difficulty in making a very accurate metallic core-box to produce a gun rifled on any system ;

little more would be required than the slide-lathe and plaining machine. The Czar's foundry would no doubt turn out a very heavy armament of rifled ordnance in this way, and that in a very short time.

Supposing a rifled 32-pounder strong enough to project an elongated shot of, say, 50 lb., with the ordinary initial velocity of a round shot, it would strike the mark if not quite close with a velocity exceeding that of the 32-lb. spherical shot, and therefore its momentum would exceed that of the 32-lb. shot more than in the proportion of 50 to 32. Such a gun would, I think, at 300 yards, easily pierce 4-in. plate. It is impossible to calculate very exactly the additional thickness which would be required to resist it, but 1½ in. would, I think, be barely sufficient. You may be opposed, however, by more powerful guns.

If the newspapers are correct and your vessels are to be of 2,000 tons burden, there is an ample margin of buoyancy, and, if difficult to add iron, perhaps oak beams would answer as a lining. Shells would not pierce the iron plates because the momentum would be diffused over so large a space, in fact there would be so large a piece to be punched out ; besides, the effective momentum would be less, or there would be two impacts, one from the materials of the preceding hemisphere of the shell, the other from the materials of the following hemisphere. Elongated shot, slightly red hot, would, I think, pass through, but whether there would be anything to be apprehended from their lodging in the oak you know much better than I do. Upon this subject, of course, I feel a great interest ; but, if possible, I feel in addition a *personal* interest. The beginning of July I wrote to the Duke of Newcastle drawing his attention to my letters to Sir John Burgoyne, and stating to him that Sir John so far agreed with me that he was of opinion that an iron target should at once be made. I mentioned to him at the same time that the preliminary experiments might be made in a week. What the Duke did I know not, but, hearing nothing, and thinking that nothing would be done, at length I sent for iron to make a target, and it had scarcely arrived when I saw in the papers the account of your experiments at Portsmouth. Before I wrote to Sir John Burgoyne I conversed with a few eminent engineers to ascertain whether they could start any insurmountable difficulty, and I found they could not. I scarcely spoke to anyone else, still I am sure if anything went wrong my name would be brought out in connection with it. It is, therefore, that I say I am personally interested. Having so far as was in my power put you on your guard, I do not wish to give you the trouble of writing to me, knowing how busy you are ; but, should you write, do not tell me any State secrets. If you did, I should be tongue-tied in the House of Lords, where, possibly hereafter, I may find some opportunity of being of use.

Believe me to be,

Dear Sir Baldwin,

Very truly yours,

(Signed) ROSSE.

Sir Baldwin Walker.

LETTER VIII.

TO SIR BALDWIN WALKER.

Castle, Parsonstown, February 23, 1855.

DEAR SIR BALDWIN,—I return the trace with many thanks. With your light deck the question of stability appears to me to present no difficulties. I perceive, however, in the new edition of Sir Howard Douglas's book, which I have just received, that he expresses doubts upon the subject ; however, on looking into the matter after refreshing my memory from my old College book, Poisson's "Mechanics," I do not see any grounds for Sir Howard's apprehensions. The centre of gravity is high, but the metacentre is much higher, owing to the form of the ship. The armament will of course, raise the centre of gravity, but I presume the engine and stores will produce nearly an equal effect, and in the opposite direction. If, indeed, there were masts of any considerable weight, and a great wave was to break upon the deck, retained for a moment by the bulwarks, while another wave threw the vessel over on the side, it is possible that the position of unstable equilibrium might be reached, and the vessel turn right over ; therefore I should think it would be unwise to employ masts and sails except the lightest possible, and it would obviously be prudent to provide the most ample means for the instantaneous escape of water from the deck. In the case I calculated roughly last spring, the deck was to be very heavy, proof against shells *thrown from mortars*, and plunging fire of any kind, and there some care would have been required to provide for the necessary stability. As to strength, I fear, in this respect, you will not find the floating battery all you could wish. From the account Sir Howard Douglas gives of the effect of shot on granite at Bomarsund, it seems almost a waste of ammunition to fire at a greater range than 400 yards, and, at that distance, I do not think the sides of the floating batteries will withstand the heavy solid shot guns with which it is said the Russians are amply provided, to say nothing of the rifling or other contrivances for projecting elongated shot. The plates, I fear, are of insufficient strength, and at the uncovered butt joints the construction is obviously weak : a shot striking within 3 or 4 in. of a joint would find a rent made for it which it would have to make were it to strike the solid plate ; and at the angle of a plate it would encounter still less resistance. To strengthen the sides materially with the present form and displacement would, I apprehend, be impracticable. I do not, however, see any reason why the bow of each boat might not be fortified by an additional layer of plates, or by other means, so as to make that part at least secure. There two of the heaviest solid shot guns might be placed, and as there will be five floating batteries, there would be ten guns to effect a breach ; an ample number, I should think, judging from the data in "Jone's Sieges," and in the "Aide Mémoire." With the bow directed to the enemy's guns, the deck, no doubt, would be much exposed to plunging fire ; but something considerable might be gained by loading the vessel at the stern so as to diminish the angle at which the projectile would strike. We have no data I am aware of sufficient to make it practicable to calculate at what angle a shot of a given size, and with a given velocity, would penetrate a wooden deck 8 in. thick. The yielding nature of the material would affect the result, and I think considerably in favour of the resistance offered by the deck. The timber would be furrowed, and the direction of the shot changed gradually, so that the deck would not at any one moment be exposed to the same violent strain as if the direction of the shot was changed *per saltum*. I do not recollect to have seen any attempt to investigate the effect of shot on timber at varying angles, either in the "Aide Mémoire," gunnery books, or the treatises on mechanics, but, as far as I can judge, I do not think there is much reason to apprehend that the deck will be penetrated by plunging fire unless the angle is considerable : it will be injured whenever struck, but I presume there will be with the plant an ample supply of planks, and spikes, to make defects good at once.

I need not add that, of course, heavy shells *thrown from mortars* will penetrate such a deck, and perhaps pass through everything ; but you know better than I do what chance there may be of the deck being so struck. Mortars can be so easily cast, ready for use the moment the vent is drilled, that the Czar may have provided an immense number of them. Pray excuse these trifling remarks : had I been equal to outdoor work, which is not yet the case, I should have been glad to have tried a few experiments so as to have been able to speak somewhat less vaguely as to the endurance of your decks under plunging fire.

Believe me to be,

Dear Sir Baldwin,

Yours truly,

ROSSE.

LETTER IX.

TO SIR JOHN BURGOYNE.

14, *Adelaide Crescent, Brighton, September* 15, 1865.

DEAR SIR JOHN.—You have so often reminded me of our correspondence at the beginning of the Russian war, about the project of casing ships with iron, that perhaps it may interest you to know the sequel. As I thought that nothing but iron could cope with granite, and as you did not suggest any insurmountable difficulties ; as, moreover, the engineers I had an opportunity of speaking to, however sceptical at first, gradually seemed to come round, I thought it might be useful to give a hint on the subject to the authorities. Mr. Nassau Senior was a great friend of Lord Lansdowne, who was a member of the Government. Moreover, Lord Landsowne had not the same prejudices against science and everything belonging to it, as official personages usually and at that time, who believed nothing which was not the result of pure practice. It appeared to me that it might be well to mention the subject to Mr. Nassau Senior, with the view to his drawing Lord Lansdowne's attention to it. Mr. Senior at once entered into the subject with his usual energy ; but he did not think that Lord Lansdowne, at his age, and without professional prestige, would be able to effect anything. He advised me to raise a discussion in the House of Lords ; my objection, however, to that was, I should only give the Russians notice to prepare bigger guns. He said the next best thing would be to communicate with the Duke of Newcastle, who, though unacquainted with such subjects, was a man of great energy, and that he perhaps would put some pressure on the Lords of the Admiralty ; that the Lords of the Admiralty, though thoroughly conversant with routine business, had not the general knowledge necessary to enable them to deal with new subjects ; and that if I wrote to them, they would merely throw the letter into the waste-paper basket. I wrote, accordingly, to the Duke of Newcastle, stating that I had discussed the question with the most competent men, and that I had no doubt as to the results of calculation : first, that 5-in. plates would be amply sufficient to cope with the Russian guns, and, second, that vessels of not very great size would safely carry them. The Duke, no doubt, sent my letter to the Admiralty, and, I presume, it speedily found its way into the waste-paper basket. It would have been wiser, as it turned out, to have acted on Senior's first suggestion, and, through the House of Lords, to have endeavoured to stir up the officials, and to have called upon the Admiralty to give reasons for their inaction. Vessels might have been constructed in time, which would have entered Sebastopol, and taken the batteries in reverse. As to the future, you seem to think the guns will beat the ships. There is a limit, however, in the nature of things, to the power of guns, and that limit, I think, will be reached sooner than many expect.

The last time Mr. Nassau Senior was with us, we talked these questions over. He was a man of very clear mind, but not a mathematician or mechanic. He was anxious to obtain information for his friend Sir G. Lewis. The question seemed to me to take this form : What is the most powerful gun which men will be able to construct of existing material ? Certainly a very general question. A formula sufficient to answer that question under all circumstances was soon obviated. The first result was that the projectile should be spherical ; then, given the minimum initial velocity which could give a sufficiently good trajectory, and given the pressure per square inch within the interior of the chamber which the material of the gun could bear without being crushed, it was easy to calculate the diameter of the largest gun which would stand. Before I gave a statement of the principle to Mr. Senior, I submitted the investigation to Mr. Purser, a very able mathematician, now Professor of Mathematics in the Queen's University, then private tutor to my son. He had the subject a week before him, and then gave in his full adhesion. I can have no doubt, therefore, that all is right. Indeed, from what I read of the Shoeburyness experiments, I believe practical men are beginning to feel what calculation had predicted, that we are approaching the limits which the nature of material has set to the power of guns. As to ships, provided that no greater height above water is insisted upon than that of the *Scorpion* or *Wyvern*, I think that we may easily plate them so as to defy the largest guns. We must have beam in proportion to the thickness of the plate, and we must have great length, if we require much speed, but not otherwise. I have great faith in the power of calculation, if we proceed cautiously ; and I believe there is but little of the information which was obtained at Shoeburyness, which might not have been previously worked out on purely theoretical considerations. I met Whitworth at Portsmouth, during the visit of the French fleet. He said they had now a prospect of obtaining better material for guns, that may perhaps extend a little the boundary line between the possible and the impossible ; but it can in no way invalidate the main result to which calculation points.

Believe me to be, dear Sir John,

Truly yours,

ROSSE.

LETTER X.

From SIR JOHN BURGOYNE.

Brighton, September 16, 1865.

MY DEAR LORD ROSSE,—You have evidently given much consideration to the very arduous problems connected with armour-plating for men-of-war ; and I know no one better able to form decisive judgments on them than one of your scientific, mechanical, and searching mind. To me, however, it is still full of difficulties, which I am quite unable to reconcile. First, to provide a covering that should be shot proof, entailing such a great additional weight to the structure of the ship as shall much lessen the capability of carrying its other necessary appurtenances of guns, ammunition, fuel, provisions, and stores, etc. Then the distribution of that extra weight round its contour, so different from the principles of diffusing a load for a good sea-going ship, hitherto thought best, are problems for nautical consideration that do not appear to be solved up to this time, except perhaps partially by having ships of enormous size.

Then you rely much more than I do on the power of the 4-in. or 5-in. plating, backed with timber, to resist shots from guns that there is no doubt can be easily manufactured. What is called invulnerable is that which is not perforated by a single or a few dispersed shot ; but *my* requirement would be the withstanding a degree of *battering* by a number.

I do not presume to take up the subject of the contest of ship against ship in all its phases, which I leave to naval men ; my reflections have been turned entirely to shore batteries as opposed to ironclads, and in particular to the defence of entrances to harbours, to rivers, etc., of moderate width—say, where the channels for the vessels do not exceed from 1,000 to 1,500 yards from the shore.

On the land, then, there will be probably 20 to 200 guns of the best of the day (according to the importance of the station), bearing on the approach, and on the passage itself ; they will be dispersed if possible, and the more they can be on elevated sites, up to 70 or 100 ft. above the water, the better, as the decks of the vessels, which are imperfectly protected, will then be more exposed than the sides. As the vessels approach end on, they will afford a considerable and very fair target to the batteries, even at great distances ; the part exposed being the decks, even from low batteries, the firing from which will be at an elevation. When entering the passage, every shot *must* tell ; for as the ship cannot run its own length during the time of the shot's flight—the guns being all laid on a concentrated part, and point blank, and at a well-known elevation, and fired when the bow is on the line of sight—she must be hit by every shot ; and if fired by signal in salvos, it may be conceived what will be the effect.

With regard to the fire of the ironclads on the batteries, it may have a powerful effect upon one of exposed masonry, insulated and compact, when opposed for a considerable time to dispersed vessels with powerful guns ; but where the shore batteries can be more or less separated in distance and height, with no exposure but of earthen parapets, or rather of their merlons and embrasures, the scattered shot directed on them in action will be thoroughly ineffective. Thus, instead of the heavy broadsides of the old men-of-war, that overwhelmed any insulated battery that might be nearly the same level, and with which it could at all close, the effect of the three or four guns, however heavy, of the ironclads, will make but little impression.

As to the invulnerability of the iron plating, it has been tried in the way most favourable to it—that is, on targets made with peculiar care, expressly for the trial, every joint and bolt newly applied and fastened, and without being previously subject to wear and tear, to the violent strains of working at sea, or to partial concussions ; and yet the $4\frac{1}{2}$ in., with its backing, has been thoroughly perforated at 200 yards range, with a Whitworth 70-pounder. It was also to be observed that, where a heavy shot made only an indent of an inch or two on the surface, the shock had driven in bolts and splinters from the timber lining within, precisely on the principle that, in the game of croquet, a ball held down to the ground firmly with the foot, and struck with violence, will drive another that is in contact with it, but loose, to a considerable distance. Thus it may be inferred that a number of hard blows from heavy shot, without absolutely penetrating, will disintegrate the entire mass, and reduce the whole to a great state of weakness.

Nor is it necessary, as I conceive, to have recourse to the enormous class of guns that are from time to time progressibly suggested. It appears to me that it would be most advantageous to employ the smallest guns that will really make effective impression, and to multiply them, in preference to having a small number of the monsters. I conceive that a piece carrying a shot of 100 to 150 lb. would have sufficient effect for the purpose.

That such guns, having a power of lancing their projectiles with high initial velocity and accuracy, as well as being thoroughly durable, might be readily manu-factured, there can be no doubt ; they would be, of course, much more easily transported, handled, and served, and by their increased numbers would have fully as great an effect—on the principle that it has been found by experiment, that two 12 or 18-pounders would breach a stone wall with the same aggregate weight of shot as effectually as one 24 or 36-pounder. I have propounded this principle as worthy of consideration, under the difficulty of manufacturing and serving enormous 300 and 600-pounders ; but the suggestion has been rejected by the Ordnance Select Committee.

Lastly, as against the power of ironclads to force entrances into harbours or other restricted passages, we have the system of torpedoes to have recourse to, which may be even rapidly improvised to a great extent with very moderate means, and with which it would be most difficult to contend.

While I argue somewhat strongly on the difficulties in the way of the ironclads performing all the great services contemplated by many enthusiasts in their favour, I am quite sensible of certain advantages they possess as opposed to other ships, which render it impossible for us not to adopt them if other maritime powers do, and not to strive how in the greatest degree to remedy the many disadvantages which as yet attend those of all countries that have studied and tried them.

My dear Lord Rosse,

Yours faithfully,

J. F BURGOYNE.

Further references to the subject : "Life and correspondence of Sir J. Burgoyne," by Wrottesley. Vol. ii. pp. 354 to 359. Correspondence between Sir John Burgoyne and I. Brunel on armour-plating, shore batteries, etc.